Structural
Failure

Structural Failure

Technical, Legal and Insurance Aspects

Proceedings of the Founding
Symposium of the International Society
for Technology, Law and Insurance

18–19 November, 1993

Vienna, Austria

Edited by

H.P.Rossmanith

Technical University of Vienna, Vienna, Austria
and Secretary-General, International Society for Technology,
Law and Insurance

E & FN SPON
An Imprint of Chapman & Hall

London · Glasgow · Weinheim · New York · Tokyo · Melbourne · Madras

Published by E & FN Spon, an imprint of Chapman & Hall, 2–6 Boundary Row, London SE1 8HN, UK

Chapman & Hall, 2–6 Boundary Row, London SE1 8HN, UK

Blackie Academic & Professional, Wester Cleddens Road, Bishopbriggs, Glasgow G64 2NZ, UK

Chapman & Hall GmbH, Pappelallee 3, 69469 Weinheim, Germany

Chapman & Hall USA, 115 Fifth Avenue, New York, NY 10003, USA

Chapman & Hall Japan, ITP-Japan, Kyowa Building, 3F, 2-2-1 Hirakawacho, Chiyoda-ku, Tokyo 102, Japan

Chapman & Hall Australia, 102 Dodds Street, South Melbourne, Victoria 3205, Australia

Chapman & Hall India, R. Seshadri, 32 Second Main Road, CIT East, Madras 600 035, India

First edition 1996

© 1996 E & FN Spon

Printed in Great Britain by St Edmundsbury Press, Bury St Edmunds, Suffolk

ISBN 0 412 20710 4

A catalogue record for this book is available from the British Library

Publisher's note
This book has been produced from camera ready copy by the individual contributors.

∞ Printed on acid-free text paper, manufactured in accordance with ANSI/NISO Z39.48-1992 (Permanence of Paper).

Contents

Preface

At the end of the 20th century the political and economic world is characterized by political nationalization and ensuing economic integration with the ultimate goal of wide internationalization of the global market place.

The momentum responsible for the development and formation of a global society characterized by extensive use of high level technology and codes based predominantly on Western ideals has become the characteristic trademark of the last decades. Development and progress are fuelled by the drive to achieve common goals and the joint performance of tasks. These tasks include the following imperatives:

- increase competitiveness
- attain a more economic attitude
- waste less energy
- increase consciousness about protection of the environment
- use resources more appropriately
- increase requirements for safety and health
- prevent failures, etc.

The majority of the complex problems associated with the preceding tasks and goals can only be solved by the joint efforts of experts from the technical, legal and insurance fields.

The occurrence of large scale and frequent structural failures has plagued industry for many years. However, only now in this highly technological age where the probability and the chances of suffering from a failure have greatly increased has the issue of economic loss due to structural failure begun to play a decisive role in economics and sometimes even in politics.

Recent detailed investigations on the economic cost of structural failures have unveiled frightening figures of 4–10% of the gross national product of the USA and the European Union. Billions of whatever currency wasted in the course of a structural failure could more appropriately be invested than in paying for loss of material and possibly human life.

Preventing failures is imperative. Currently, some of the most pressing issues are concerned with ageing of aircraft and bridges, plant life extension in technologically less advanced countries, passenger mass transport, and redesign and increase in safety of nuclear power plants. In many cases a better understanding of material behaviour will lead to improved design and hence will increase the operational and safety reliability of the structural component or of an entire plant. Results of an integrity evaluation of a plant serve as a basis for economically oriented decision making.

Failures of technical reliability will call for legal liability. A severe design fault which renders a product harmful may by means of a failure case shift the scene of action from the technical field into the legal and insurance fields. Harmonisation of product liability laws is sought in the European Union as well as worldwide in order to tame the radicalisation movement that is known to exist in the USA.

International engineering disputes in the aftermath of a large scale structural failure call forensic experts to work and expert opinions are displayed, assessed and judged, eventually in court. Then, issues of licensed protective rights, patents and patentability, technical documentation and abuse of user instructions, standards, software qualification and many other topics may become decisive. Recent investigations in various fields of engineering have shown that the majority of failure cases have their primary reason in human error and human failure. Thus, education and training of personnel is a key issue in the prevention of failures.

The many large scale failures associated with chemical plants, the oil producing industry, and tanker accidents have led to an increasing awareness by the public. Environmental protection and liability for environmental damage have added a new dimension to the issue of structural failure. The risk-conscious society at the dawn of the 21st century now focuses much more attention on social, psychological and moral effects of fatalities as a result of a large scale structural failure.

ISTLI, the International Society for Technology, Law and Insurance has been formed with the aim of establishing a common basis and platform for the interaction and cooperation between engineers, scientists, lawyers, managers and technical insurers within the framework of structural failures. ISTLI is proud to be the first interdisciplinary, inter professional and international society addressing technical, legal, insurance, economic, ethical and moral issues associated with structural failures.

H.P. Rossmanith
ISTLI Secretary General
Vienna

Foreword

W. SCHUPPICH
President of DACH

On behalf of the European Association of Lawyers (DACH) and the Presidential Conference of European Lawyer's Organisations I would like to welcome the distinguished guests and participants to the Foundation Meeting of the International Society for Technology, Law and Insurance. It is my pleasure to see ISTLI based in Vienna and spreading its activities from Austria. Austria, located at the Western fringe of Eastern Europe and the Eastern fringe of Western Europe has always served as a go-between for the integrating European industrial societies and their Eastern European neighbours.

The tasks and objectives of ISTLI are fascinating. Perhaps not obvious at the first glance, interdependencies of a practical and scientific nature do exist and will be established between

- Technology as the scientifically based exploitation of natural forces for the satisfaction of mutual and joint requirements and for the removal of existing obstacles,
- Law as the science of an ordered balance between countercurrent interests and the mastering of inter-human disagreements, and
- Insurance, as the mediator of all these fields reducing the risks involved and offering new perspectives and chances.

The phenomena leading to the welding of the three fields are to be found in a safer economic and social environment. With the increasing improvement of social conditions in our science- and technology-based civilisation the calls for a safer and less risky way of life persists. The degree of subjective risk consciousness is counter-parallel to the reduction of risk and objective increase of public safety. A growing fear of the unknown danger can be noticed.

The more the law enters our life the less we find a clear path in this labyrinth of legal issues. Technical knowledge about risks involved, ie more detailed perception and in-depth understanding and improved adaption of the body of laws according to the actual state-of-the-art of science and technology, leads to an ever-increasing flood of the cry for safety and legal protection to comply with changed social environment, the larger the requirement for safety and economic measures. Guidelines offer a basis for conspicuous subjective safety by prescribing predictable modes of conduct. Lack of legal order introduces the worries of anarchy and danger. It is therefore necessary to reconsider and assess the body of regulations and standards with respect to continuous assurance of legal safety given the state-of-the-art of technology. Adequate protection may be offered by insurance companies, which know how to fight against the bad effects of new technology.

Risk defense and technology defense, as well as law practice, have gone international and are basically controlled by supply and demand market mechanisms.

In each of my functions regarding improving the harmony of law practice in Europe, I have been and I am happy to see ISTLI operating on a global scale by building bridges not only between countries and professional groups. In addition, Vienna has been recognized as the European centre of lawyers. As seen from the viewpoint of Austrian advocates, ISTLI's aims and goals have extraordinary meaning and importance.

Contributors

C.-O. Bauer
President of ISTLI
HDI Sicherheitstechnik GmbH
Riethorst 2
D-30659 Hannover
Germany

M. Bily
Institute of Materials and Machine
      Mechanics
Slovak Academy of Sciences
Bratislava
Slovak Republic

A. Bussiba
Nuclear Research Center - Negev
PO Box 9001
Beer-Sheva 84190
Israel

Krzysztaf Czerkas
Failure Analysis Associates
Sp z.O.O.
ul. Uphagena 27
80-237 Gdańsk
Poland

E. Corazza
Institute for Research and Testing in
      Materials Technology (TVFA)
Technical University Vienna
Karlsplatz 13
A-1040 Vienna
Austria

M. Drdacky
Academy of Sciences
Institute of Theoretical & Applied
      Mechanics
Vysehradska 49
128 49 Prague 49
Czech Republic

L. Faria
Sociedade Portugesa de Materiais
Instituto Superior Tecnico
Av Rovisco Pais
1096 Lisbon Codex
Portugal

H. A. Gomide
Dep. Engenharia Mecanica
Universidade Federal de Uberlandia
Rua Duque de Caxias
285 Caixa Postal, 593
38400 Uberlandia MG
Brazil

M. Itabashi
Science University of Tokyo
Dept of Materials Science & Technology
2641 Yamazaki
Noda, Chiba 278
Japan

Y. Katz
Nuclear Research Center - Negev
PO Box 9001
Beer-Sheva 84190
Israel

K. Kawata
Science University of Tokyo
Dept of Materials Science & Technology
2641 Yamazaki
Noda, Chiba 278
Japan

R. Kieselbach
EMPA Schadensanalytik Metalle
Überlandstrasse 129
CH-8600 Dübendorf
Switzerland

J. Scott Kirkwood
Professor of Business and Economics
Wingate College
Wingate
NC 28174
USA

S. Kusaka
Graduate School of Science
University of Tokyo
2641 Yamazaki
Noda, Chiba 278
Japan

I. Le May
Metallurgical Consultants Services Ltd
PO Box 5006
Saskatoon
Sask. S7K 4E3
Canada

S. N. Mannheimer
Attorney at Law
Rio de Janeiro
Brazil

T. Pachmann
Frick & Frick, Attorneys at Law
Uraniastrasse 12
Postfach 996
CH-8021 Zürich 1
Switzerland

R. K. Penny
R. K. Penny & Associates
PO Box 30136
Tokai 7966
South Africa

A. Puri
GEC, Marconi Defense Systems
Warren Lane
Stanmore, Middlesex HA7 4LY
United Kingdom

P. Révy von Belvard
Patent Lawyer
Büchel, von Révy & Partner
Im Zedernpark
CH-9500 Wil
Switzerland

B Ross
Failure Analysis Associates, Inc.
149 Commonwealth Drive
PO Box 3015
Menlo Park
CA 94025
USA

H. P. Rossmanith
Secretary General of ISTLI
Institute of Mechanics
Technical University Vienna
Wiedner Hauptstrasse 8-10/325
A-1040 Vienna
Austria

V. I. Schevchenko
Volgograd Civil Engineering Institute
Volgograd
Russia

F. Schwank
Attorney at Law
Borsegebaude
Wipplingerstrasse 34
A-1010 Vienna
Austria

T. L. da Silveira
Associacao Brasileira de Ciencias
    Mecanicas
Av Rio Branceo, 124-8 andar
20040-001 Rio de Janeiro, RJ
Brazil

Dr E. Urbańska-Galewka
Technical University of Gdańsk
Civil Engineering Faculty
ul.G.Narutowicza 11/12
80-952 Gdańsk
Poland

T. Varga
Institute for Research and Testing in
    Materials Technology (TVFA)
Technical University of Vienna
Karlsplatz 13
A-1040 Vienna
Austria

T. Yamada
Science University of Tokyo
School of Management
Kuki Campus
Shimokiyo-ku
Saitama 346
Japan

Yu V. Zaitsev
Moscow State Open University
Moscow
Russia

1 ECONOMIC EFFECTS OF RUIN OF SYSTEMS

LUCIANO FARIA
Instituto Superior Técnico, Lisbon, Portugal

Abstract
Although a continuous effort in R&D had been made toward improved understanding of the behaviour of materials, structures and products, development of materials more resistant to fracture and fatigue, to corrosion, to erosion and wear, improved processing techniques, utilization of advanced design concepts, advancement of equipment for inspection and preventive maintenance, development of proper utilization of products by users, ruins and even catastrophic failures could occur in systems, causing important losses to the economy.

American and European studies on the economic effects of fracture, corrosion, erosion and wear lead to the conclusion that these global losses attain about 10% of the GNP of the USA and EC per year.

Factors contributing to fracture and other causes of ruin are presented and means to reduce the corresponding costs to about 50% are indicated.
Keywords: Fracture, fatigue, corrosion, erosion, wear, economic effects.

1 Introduction

Materials, structures and other products made from them are all subject to ruin, due to different causes, mainly due to fracture, corrosion, erosion and wear.

Substantial effort is continuously directed toward improved understanding of the behaviour of materials, structures and products, development of more resistant materials to fracture and fatigue, to corrosion, to erosion and wear, improved processing techniques, utilization of advanced design concepts, advancement of equipment for inspection and preventive maintenance, development of proper utilization of products by the users.

In spite of these efforts and research it is still necessary to overdesign the systems to assure their reliability and to prevent ruins and even catastrophic failures.

The overdesign, the ruins and their consequential effects are the origin of important losses for the economy of every sector and every country and represent uses of resources that could be diverted to other worthwhile purposes.

These losses justify the realisation of studies and of research leading to their

Structural Failure: Technical, Legal and Insurance Aspects. Edited by H.P. Rossmanith. Published 1996 by E & FN Spon, 2–6 Boundary Row, London SE1 8HN, UK. ISBN: 0 419 20710 4.

evaluation and to find the proper solution to minimise them.

2   North-American Studies

The National Bureau of Standards and the Battelle Columbus Laboratories made in 1978 a study to assess the costs of material fracture to the United States and therefore their influence to the American economy. The objectives of this study were [1]:

I.   To assess the total cost of fracture in the US economy. This total includes the costs associated with the occurrence of unintended fracture and the costs associated with the prevention of fracture.

II.   To assess presently reducible costs of fracture. The costs of fracture could be reduced if economically best fracture control practices were to be universally applied throughout the economy. This reduction in cost can be approached through transfer of known technology.

III. To assess the future reducible costs of fracture. The costs of fracture could be reduced in the future by conducting research to advance the understanding of material fracture and fracture control technology.

This study considered the effects in 154 industrial sectors and concluded that the largest sectors contributing to the costs of fracture were [1]:

- motor vehicles and parts (1),
- aircraft and parts (1),
- construction, residences and non-residential buildings (2),
- all other food and kindred products (2),
- all other fabricated structural products (1),
- all other non-ferrous metals (1),
- petroleum refining and related products (1),
- structural metal (1),
- tyres and inner tubes (2).

The number (1) indicates the sectors that utilise mainly metals, and the number (2) indicates those that utilise other kinds of materials, such as concrete, polymers, wood and derivatives.

The total costs of these economic effects attained 4% of USA GNP, that means $119 billion per year (1982 dollars); the estimated uncertainty was $\pm$ 10%.

The factors that contributed most significantly to the costs of fracture were the following [1]:

- The imposition of large factors of safety in many structures reflects uncertainty in design and results in increased expenditures for construction, materials production, and transportation.

- Many industries devote significant resources to repair and maintenance, replacement, and scrapping associated with equipment failures.
- Inspection with respect to material quality control and structural reliability consumes significant resources.

The analysis of the results of this study lead to the following conclusions:

- in the short term, fracture costs could be reduced by about 30% (1.2% GNP of USA) by application throughout the economy of the best fracture control (design, manufacturing, inspection, maintenance, repair) and transfer of advanced technology;
- in the medium term, research directed toward fracture related problems (material and structural) could further reduce the costs of fracture by about 24% (1% GNP of USA);
- the reminder of the costs of fracture - 46% (1.8 GNP of USA) were not considered reducible at the time of the study.

An earlier study lead by the same entities on the "cost of corrosion" arrived practically at the same results: the economic effects on the USA economy due to corrosion attain about 4% of the USA GNP. The largest sectors that contribute to these losses are mainly the following:

- motor vehicles, aircraft and parts,
- nuclear and electronic industries,
- petroleum refining,
- structural metal,
- paper and beer industries,
- mineral ore industry,
- metallic implants
- chemical and pharmaceutical products,
- batteries and combustible cells.

3   European studies

The European Community considered this American study of a paramount importance and decided to make a similar study about the "economic effects of fracture in Europe". The European Joint Research Center of Petten took the initiative for this study and committed the ESIS (European Structural Integrity Society) to make it. Professor Luciano Faria, president of the ESIS Strengthening and Liaison Committee, was appointed as project leader and rapporteur.

The objectives of the European study were similar to those of the American study and the report presented an economic evaluation of the effects of fracture for each type of equipment and for each industrial sector. The report presented some recommendations for future work concerning research, improvement of codes and standards, training and education of skilled people and engineers, and suggestions for effective transfer of know-how from researchers to industrial companies.

This study, although the adopted methodology is different from that adopted by the American study, lead practically to the same conclusions: the losses of the 12 European countries belonging to the European Community, due to fracture, attain about 4% of the global GNP with an estimated uncertainty of $\pm$ 25%.

An earlier European study concerning the losses due to corrosion lead also to similar conclusions of the American study referred to above. If we consider the effects due to erosion and wear, their costs according to British and German studies attain 2% of the GNP of these countries.

4   Factors contributing to fracture

Taking into consideration the American and European studies and the information given by insurance companies, the factors that contribute mainly to fracture are the following, the average influence being indicated between brackets:

- product defects on failures:
  planning faults and design faults (including incorrect calculation) (15%),
  manufacturing and repair faults (13%),
  material faults (5%),
- operating faults and maintenance errors (45%),
- assembly, erection and installation faults (12%),

The external and unknown causes attain about 10%.

Besides the above mentioned factors, if we also consider indirect and consequential costs, the influence of the cost of fracture on the total costs of each product varies between 25% (for heavy equipment with high consequential costs) and 2% (for medium size and light equipment with low consequential costs). But in certain well-known cases the cost of fracture attained 100% and more of the cost of the product.

The indirect costs derive mainly from the following factors:

- uncertainty in design and in evaluating stresses and strains;
- poorly understood behaviour of the materials used in manufacturing;
- increase in weight due to the imposition of large factors of safety;
- influence of manufacturing process on the behaviour of the systems: introduction of defects, residual stresses, etc;
- damage due to transportation or unsuitable handling;
- expensive quality control, or quality control equipment not detecting all existing defects;
- repair procedures introducing supplementary causes of failure.

The consequential costs derive from the following aspects:

- fatalities, immediate or delayed;

- injuries, disabling or non-disabling;
- shock, individual or societal/social;
- effects to mental health, long or short term;
- disruption to people's way of life;
- environmental damage, pollution;
- financial losses: property damage, stop in production, increase in stock of spare parts, other causes.

5 Reducing the cost of fracture

Considering the average influence of the different factors, it seems possible to reduce the influence of the cost of fracture on the cost of each product by about 50%, by means of:

- improving codes, standards and design rules,
- training designers and skilled people,
- transfer of existing knowledge developed by researchers,
- introducing into companies total quality rules.

These means could assure better reliability, better quality, better competitiveness.

Research is needed to improve codes and standards for better design rules, for better knowledge of the behaviour of existing and new materials, in present and in new applications, developing testing standards in areas where practical standards do not exist, developing fracture mechanics.

The potential savings in Europe attain in our opinion the same percentage as indicated in the American study, that is 2% of the GNP (8×10^{10} 1991 ECU), if the above suggested measures are adopted.

6 Reducing the cost of ruins

It seems also possible in our opinion to reduce the economic effects of ruins to corrosion, erosion and in the same percentage of the reductions referred to for fracture ruins, if similar measures are taken.

Therefore it is apparent that the material durability costs are substantial and attain about 10% of the GNP of the USA and EC, if we consider the effects due to fracture, corrosion, erosion and wear.

The required action on education and training, knowledge transfer and R&D should be planned for a certain period, not less than 10 years, in order to get the expected results. The financial effort required is very small if we compare it with the losses that could be eliminated; we estimated the cost of these actions in about 0.3% of these savings. Investing in R&D and in knowledge transfer for better structural integrity could therefore be one of the best investments ever made.

7 References

[1] "The Economic Effects of Fracture in the United States", *US Department of Commerce, National Bureau of Standards, Special Publication 647-1*, March 1983.
[2] "The Economic Effects of Fracture in the United States", *US Department of Commerce, National Bureau of Standards, Special Publication 647-2*, March 1983.
[3] "Corrosion Economic Calculations", by E D Verink, Jr, in *Metals Handbook, Vol 13 - Corrosion*, ASM International, Metals Park, Ohio, 1987.
[4] "The Economic Effects of Fracture in Europe - Final Report", *Commission of the European Communities*, © 1991.
[5] "Economic Effects of Fracture", by Luciano Faria, *Third International Conference on Structural Failure, Product Liability and Technical Insurance*, Vienna, July 1989, *Forensic Engineering - The International Journal*, Volume 2, Number 1/2, 1990, Pergamon Press.
[6] "Allianz Handbook of Loss Prevention", VDI Verlag, 1987.

2 DESIGNING AGAINST FAILURE

ROY K. PENNY

R.K. Penny and Associates, Tokai, Cape Town, South Africa

Abstract

This paper reviews the crucial importance of capturing the essential characteristics of materials through simple but robust idealisations of complex behaviour. It reveals the growing need for design against failure which is not being helped by attempts to mimic reality in the belief that larger and faster computers will relieve designers of their duties to think. This is not to say that basic design concepts cannot be aided and further developed with large computers. They can, and particularly as a cost-effective, simulation tool and as a tool for designing and conducting sensible in-plant investigations and the rational planning of maintenance procedures. The background to these approaches will have to come from a comprehensive understanding of the principles and interactions of mechanics and materials. This will aid decisions that have to be made about hazard prediction, damage and its repair and life extension schemes.

The philosophy expressed in the paper is not specific to a particular technology or to operation in a particular part of the world. Rather, it aims to show, by example, a non-specialist approach to mechanics related materials problems involved in typical design situations.

The needs for more purposeful schemes of technical education and goal driven research are central to all initiatives for a renewed emphasis on failure prevention and damage assessment.

1 Introduction

Everything is made of some material but it is not only the properties of materials that dictate poor performance or the failure of parts and machines. It is through the rational use of mechanics, linked with intelligent material choice – part of a process called design – which will assure success. In fact, design aided by mechanics can, should and sometimes does, encompass the development of new materials having desirable properties on a microscale ; examples are in advanced metal alloys and fibre reinforced polymers and ceramics. Similarly, "damage" can occur on a scale easily visible to the eye – say, a cracked rotor – or on the sub-microscopic level – creep cavitation for example. In both cases, the idealisations, modelling and numeracy used in mechanics allow us to capture the essential features of how parts behave rather than formulating laws governing universal reality. Metallurgical inputs based on the experience and rudimentary understanding of specialists will, of course, help on the path to the fundamental understanding of materials but it is very noticeable, over many years, that the robust methods arising out of practical needs have been, and still are far in advance of "scientific" achievement. In

Structural Failure: Technical, Legal and Insurance Aspects. Edited by H.P. Rossmanith. Published 1996 by E & FN Spon, 2–6 Boundary Row, London SE1 8HN, UK. ISBN: 0 419 20710 4.

1654 Hooke[1] was ignorant of the fact that the material samples he tested were grossly inhomogeneous in their makeup and manufacture, but we still depend heavily on the bulk results of his tests, wherein inhomogenity averaged itself out (*ut tensio sic uis*). Wöhler[2] devised his testing machine to solve problems of railway axle failures, not for the purpose of studying fatigue. If the Wright brothers had waited for the material that "... needs be a hundred-fold stronger that steel ..." or "... for the reversal of gravity ..." as necessary requirements suggested by eminent specialists in materials and mathematics of the day, it is unlikely that all of us would be at this meeting in Vienna.

These, and other apocrypha, are well known, yet the basis of their origins is continually being expanded by universities, government organisations and other bodies through what is rather loosely termed "research". This seems to stem largely from the emergence of more sophisticated machines for testing materials in the form of similar type specimens used by Hooke and Wöhler but rarely is this accompanied by goal orientation ; the apparent elegance and timelessness of research is now being aided by the use of super-computers. All this, at a time when succinct procedures for design against failure and for greater safety and economy in schemes for life extension of existing equipment and in the design of future, even more potentially hazardous machines, are not accorded the same status. Difficult and ill-defined problems have been avoided by specialists for too long. It is not unreasonable to expect progress if more talented people can turn their attentions to future needs by capturing the essential behaviour of materials through robust idealisations of complexities – rather then having the same people trying to mimic reality in the belief that faster computers will relieve designers of their responsibilities to think. This is not to say that basic design concepts cannot be aided in their further development with large computers. They can, and particularly as a cost-effective simulation tool and as a tool for designing, planning and interpreting in-plant investigations and for devising procedures of inspection and maintenance.

The background to these fresh approaches will have to come from a greater and more comprehensive understanding of the principles and interactions of mechanics and materials. This will aid in making decisions about hazard prediction, damage and its repair and life extension schemes. One such area in need of urgent attention concerns the difficult class of problems involved in the design, life assessment and life extension of plant and machinery operating at high temperature. These problems involve highly non-linear, potentially dangerous processes, such as creep, fatigue, stress corrosion and sometimes combinations of any of these. Future situations which involve viscous flow of materials at high operating temperatures and loadings need to be incorporated into the same approaches whenever possible.

The philosophy expressed in the paper is not specific to a particular technology or to operation of plant in a particular part of the world. Rather, it aims to show, by example, a non-specialist approach to some mechanics related material problems in high temperature design. The approach reviewed in the paper derives from the pioneering work in Russia[3],[4] on damage as well the "reference stress" techniques initiated in the UK[5] during the 50's and 60's.

These mechanics related materials solution procedures have started to find their way into Codes of Practice[6] and form a solid basis for wider use by designers and for better understanding of ways for failure minimisation and life extension. The purpose of this paper is to provide brief reviews of some of these techniques.

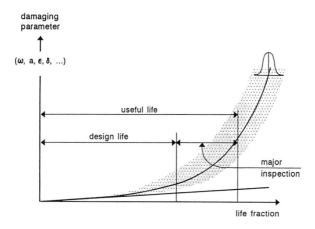

damaging
parameter

$(\omega,\ a,\ \epsilon,\ \delta,\ ...)$

useful life

design life

major
inspection

life fraction

Figure 1: Schematic representation of damaging parameters

2   Overview of a Pressing Problem Area

The problem area of present interest is that which involves continuing irreversible material damage caused by mechanical loading and environmental features which result eventually in unacceptably high rates of change in component damage. The term damage could initiate from cavity formation (ω), micro cracks (a) and be manifested in terms of strain (ε) and gross deformation (δ). Some or all of these forms of damage could be accelerated by creep at temperatures above about one third of the material melting point as well as by other processes, such as corrosion, spalling, irradiation. These features are depicted in Figure 1 together with their stochastic nature which is inevitable. Typical lifetime expectations could be minutes to decades but expressed in terms of life fractions, the form of the schematic is much the same for different situations. The designers's task is to provide for a safe life using Codes as guidelines. The operator's task is to seek a useful life by extending the design life, to do so with a minimum of interruptions in the operation and with close bounds on failure probabilities. In some cases, e.g. combined high temperatures and high loadings, Codes do not exist. In all cases, there is likely to be a shortage of (even) bulk materials behaviour data ; laboratory tests from which these data are obtained are often performed outside the regimes of component operation.

3   Some Particular Aspects of Design for Creep

The discipline of design for creep is a particularly useful one for the illustration of guidelines that are offered by mechanics related materials problems. At least five basic problem areas are involved:

3.1 Laboratory acquired bulk materials properties data is time consuming, costly and difficult to obtain without considerable scatter of results. This reveals a number of important areas where the use of mechanics can help to optimise the amount of data to be collected with respect to cost and accuracy. Amongst these are a) the specification of reliable test procedures from which maximum data are obtained, b) minimising the number of tests and c) providing a rational means for extrapolation of short term results to the prediction of longer terms.

Of these items, useful mechanics-based guidelines for creep specimen testing in terms of the specimen design, axial alignment and temperature control have been provided and successfully implemented several years ago[5]. In spite of these guidelines, the wasteful practice of conducting "rupture" tests in which full data regarding creep accumulation is not catered for is still pursued. The minimisation of specimen creep testing has been suggested on the basis of reference stress techniques[5]. These techniques are described elsewhere[5, 10] but, in essence, the minimum number of tests required is one. Of course, this is an ambitious ideal but since the origination of reference stress methods, there is sufficient evidence in favour of use of the methods for their continued development ; certainly such methods merit further goal directed research. The reference stress approach is also capable of reasonable approximations of the behaviour of components containing cracks[7] or which are subject to buckling during creep[5].

So far as extrapolation techniques are concerned, developments which apply at high stress ("Ductile") regions as well as low stress ("Brittle") levels are essential. A recent contribution towards this is described later in this paper.

3.2 Materials data acquired on laboratory specimens is not sought for their own sake but rather to predict the behaviour and reliability of components made from the same materials. Which are the most effective ways to accomplish this requirement?

3.3 In order to ensure the reliable operation of components in service, inspection and maintenance procedures must be devised in ways which inspectors know where to look and what to look for. Periods between inspections need to be maximised without jeopardising reliability. Some speculative ideas for this are given later.

3.4 Life extension schemes must cope for damage in regions where the damage rate has accelerated to the point of initiating cracks. Detection of interior cracks (at weld interfaces for example) or at surfaces, their nature and extent is essential for reliable remnant life assessment which could result in repair or component replacement.

3.5 Promising methods for defect detection have been greatly advanced recently through modern electronics and data collection schemes by computers and this is particularly true in ultrasonics, thermography and, more recently, holographic interferometry. Quantification of the discriminatory capabilities of video holography have been made through the use of mechanics[8]. Success in the use of holographic techniques leads to possible methods for on-line schemes for plant monitoring. These and other viewing and data capturing methods could be expensive since complete coverage is not likely. A choice for discrete positioning of these devices in order to make them economically viable, could be made through rational use of mechanics by assessing the most likely places of damage accumulation.

4 Developments of Mechanics Related Techniques in Design for Creep

The most significant developments in design for creep over the last few decades are acknowledged to be those involving concepts of material damage kinetics[2] and reference stresses[5]. Both of these arose from a lack of "scientific" progress of any significance to help designers charged with the huge responsibility for the integrity of plant and machines being operated at increasingly higher loading and temperature conditions. The techniques which evolved were phenomenological in approach as a result and are, nonetheless, practical and rigorous in their derivations.

In damage studies, a global parameter representing cavity formation or other complicated processes was introduced to show how rising average stresses due to the presence of damage would lead to strain rates accelerating to infinity at finite values of strain in finite times. This bold and ingenious scheme fits all the usual observations of, so-called, tertiary creep and stress/rupture time characteristics. Like Hooke, Kachanov side-stepped local, internal inhomegenieties, which he could not see or measure and focussed on bulk behaviour which is easily measured. Various research teams around the world followed the leads provided by Kachanov. The most promising of these appears to stem from the French work[9] which extended the basic ideas into areas such as creep-fatigue interactions and low cycle fatigue.

Reference stress techniques originally arose out of the need for pragmatic ways for calculating creep deformation accumulation under variable loading. At the time, even the apparently simple problem of the creep of pipes under variable pressure and temperature could not be tackled because of insufficient computer power. It is not surprising that the same is true today if one were to think in terms of, let us say, the creep analysis of a nozzle welded into a reactor vessel subject to variable pressure and temperature as well as nozzle reaction forces. Even if sufficient computing power were available there is insufficient or inappropriate materials data. Such problems can be dealt with by using reference stress techniques which aim to correlate the creep behaviour in a structure directly with an equivalent simple creep test performed at a stress level which, in some way, is representative of the structure ; hence the term reference stress. The original concept was, of course, consistent with bulk behaviour rather than local effects but has since been extended to parts containing defects. Initially, reference stresses were derived through the intuition that there should exist a scaling factor between bulk structural deformation and specimen creep strain. As a result, numerous calculations on elementary structures were made to define the stress level at which this was so. The scaling factor so found was realised to be geometry dependent and not material dependent. This led to the further realisation that the bulk behaviour of a structure is dictated by its collapse loading. Consequently the reference stress $\bar{\sigma}$ is most simply defined as :

$$\bar{\sigma} = \frac{P}{P_u}\sigma_u \tag{1}$$

where P is the structure loading and P_u and σ_u are the ultimate load carrying capacity of the structure and of a tensile specimen respectively. Such a condition must exist at any time during any degrading process : the structure and the material from which it is composed eventually become exhausted. Further advantages of this definition of the reference stress are that it is independent of precise material data and in many cases ultimate load factors are already available ; if they are not, they could be determined experimentally or in some cases by finite element methods.   70

4.1   Some Examples

Numerous examples of mechanics related problems involving creep and the damage and reference stress techniques have already been published ; references [5], [10] and [11] are typical of these but the British Code of Practice R5[6] is probably the most up-to-date in its applications. Other illustrations which follow here are speculative and are aimed at the practical problems of rupture data extrapolation, remnant life assessments based on in-plant measurements and, finally, dealing with uncertainty.

4.1.1   Extrapolation of Rupture Data

Several methods for this are already available ; examples are the θ method of Dorn and parametric methods of Larson-Miller, Manson-Haferd and others which are described in [5]. The present object is to use damage concepts in extrapolation processes and, while this has already received much attention, present methods seem to rely on a bi-linear approach wherein it is supposed that high stress regions must be separated from low stresses by the two hatched lines shown in Figure 2. Of course, it is realised that this bi-linear approach is an approximation to the full line in the Figure. It would certainly be more useful if the continuous lines, could be predicted since it may then be possible to make extrapolations from short term data more reliable. In addition it would be convenient to have a unified approach for high and low stresses for components which operate in different regions of the rupture curve.

The simple analysis which follows is based upon Figure 3 which is a schematic of the σ, ε, t variations occurring in "static" testing – the σ, ε plane, during creep measurements – the ε, t plane and finally, Kachanov type stress variations resulting from internally

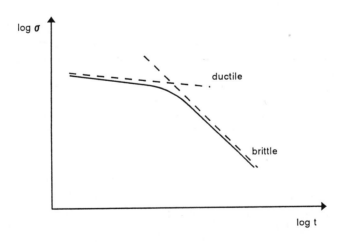

Figure 2: Stress-rupture time schematic

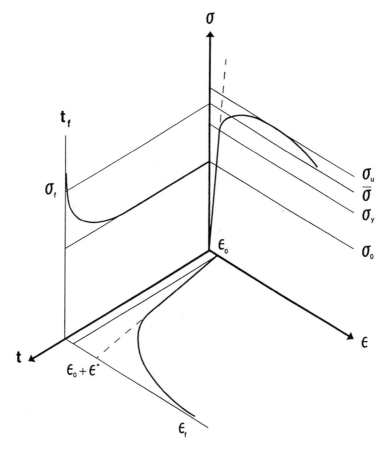

Figure 3: A stress-strain-time schematic

generated damage - the σ, t plane. Following the reference stress notion, collapse of the structure (in this case, a specimen under constant load) will occur when a critical stress level σ_f is reached at time t_f ; this can be regarded as the point of failure because of the occurrence of very high rates of strain.

Following Kachanov, the stress σ_t in a specimen under constant load at time t is

$$\sigma_t = \sigma_o \frac{A_o}{A_t} = \frac{\sigma_o}{(1 - \omega)} \tag{2}$$

where A_o is the original cross-section area and $1 - \omega$ is the area loss at time t when the area is A_t.

Departing here from Kachanov, failure will be taken to occur when $\omega = \omega_f$, which represents that loss of area giving rise to unacceptably high strain rates of the specimen for whatever reason. To determine ω_f (which is usually taken as 1) we define σ_f ($\equiv \bar{\sigma}$) from σ_t in equation 2 whilst assuming this value will be reached at a level between σ_y and σ_u in a strain-hardening material (Figure 3). Then

$$\frac{\sigma_o}{(1 - \omega_f)} = \bar{\sigma} = \sigma_y + \alpha(\sigma_u - \sigma_y) \tag{3}$$

where α is a proportion above σ_y and which will depend on the material being tested. From (3) it follows that

$$\omega_f = 1 - \frac{\sigma_o}{\bar{\sigma}} \tag{4}$$

where $\bar{\sigma} = \delta \sigma_y = \sigma_y \left[1 + \alpha \left(\frac{\sigma_u}{\sigma_y} - 1 \right) \right] \tag{5}$

The choice of δ could be made in several ways. For example, $\delta = 1$ for non-hardening materials ($\sigma_y = \sigma_u$), or $\alpha = 1$, is an extreme case. The value $\delta = m/m + 1$, where m is the Norton creep exponent is a possible bound suggested elsewhere [12] which would give highly conservative results. Thus, $m/m + 1 \leq \delta \leq \sigma_u/\sigma_y$ represents a family of possibilities for defining failure of the specimen in practical terms. In all cases, it is worth noting that $\omega_f < 1$.

Returning now to Kachanov's approach the damage rate is expressed as

$$\frac{d\omega}{dt} = \frac{B\sigma_o^k}{(1 - \omega)^r}, \quad 0 \leq \omega \leq \omega_f \tag{6}$$

where B, k and r are material constants, easily obtained by experiment. Integration of (6) gives the usual ω, t relationship

$$(1 - \omega)^{1+r} = 1 - B(1 + r)\sigma_o^k t \tag{7}$$

Fast failure occurs at time t_f when $\omega = \omega_f$ so that substituting (4) into (7) gives for the failure time t_f :

$$t_f = \beta t_R \tag{8}$$

where $t_R = \frac{1}{B(1 + r)\sigma_o^k} \tag{9}$

is the Kachanov result for "brittle" failure and

$$\beta = \left[1 - \left(\frac{\sigma_o}{\bar{\sigma}} \right)^{1+r} \right] \tag{10}$$

is a factor dictated by the level at which the test is being performed in relation to a test performed at time near zero. The form of t_f plotted in the usual (double logarithm axes) is shown in Figure 4. For most metallic construction materials it is likely that departure from the "brittle" rupture curve will not be significant up to about $\bar{\sigma}/2$ for which interval, $\beta \approx 1$. Above about $\bar{\sigma}/2$ the value of β falls rapidly (Figure 5).

In order to test the applicability of these notions, experimentally obtained rupture results have been taken from the literature[13], [14] for two alloys which cover tests on these materials at four temperatures and over times up to 100,000h. Calculated results for these tests are shown in Figures 6a, b, c and d and summarised in Figure 7. These are good enough to encourage further exploration. For example, could the methods be helpful when planning rupture testing – what levels to test at? ; at what level could longer term results be predicted from short term ones? The method outlined could also be helpful to those designing at high stress levels and/or high temperatures where viscous behaviour applies.

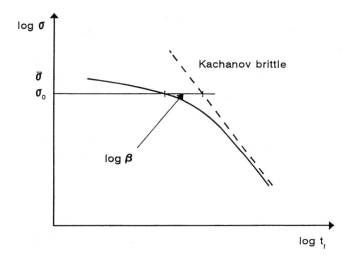

Figure 4: Characteristics of the modified Kachanov brittle rupture curve

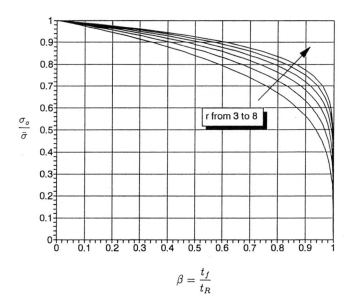

$$\beta = \frac{t_f}{t_R}$$

Figure 5: Variation of modifying factor β with stress level

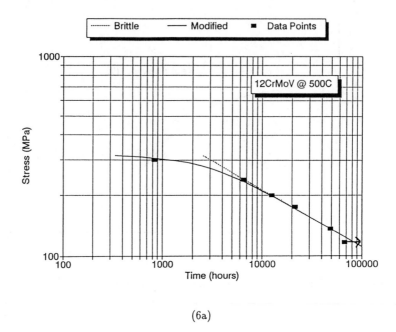

(6a)

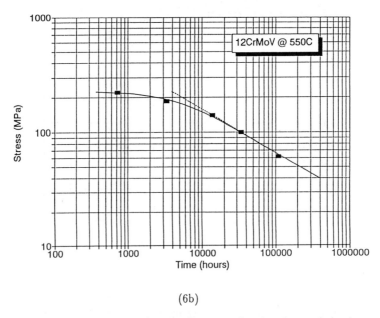

(6b)

Figure 6: Rupture curves for two alloys: predicted and recorded values

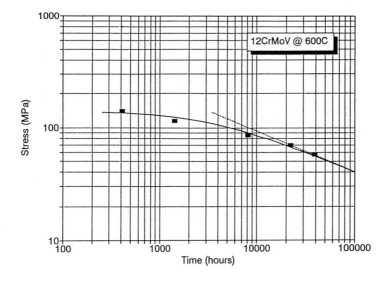

(6c)

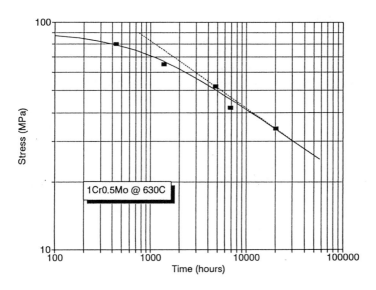

(6d)

Figure 6 continued

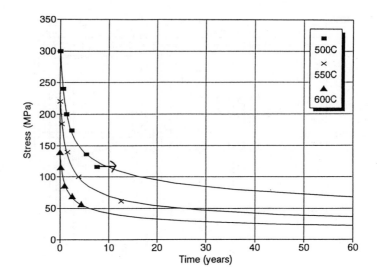

Figure 7: Stress rupture summary for 12CrMoV alloy: linear scale

4.1.2   Strain Predictions

Following on from the results of 4.1.1 which indicate some modifications to the usual damage calculations, it is worth investigating the effects of these modifications on strain predictions. The same concepts are used here ; these amount to modification of Norton's law but allowing for damage and consequent stress changes in a test performed at constant load.

Following the usual procedure [11] :

$$\dot{\epsilon} = \frac{\dot{\epsilon}_o}{(1-\omega)^p} \tag{11}$$

where (˙) represents differentiation with respect to time t if primary creep strains are neglected or with respect to the time measure τ [5] if they are not ($\tau \propto t^n$, $1/3 \le n \le 1/2$ usually).

Substituting for $1 - \omega$ from (7) into (11) gives for the current strain at time t:

$$\frac{\epsilon}{\dot{\epsilon}_o t_R} = \lambda \left[1 - \left(1 - \frac{t}{t_R} \right)^{1/\lambda} \right] \tag{12}$$

$$\text{where} \quad \lambda = \frac{1+r}{1+r-p} \tag{13}$$

Fast failure occurs at $t = t_f$ (as in Section 4.1.1 above) when $\epsilon = \epsilon_f$. This results in the following equations expressed in terms of the life fraction $\Gamma = t/t_f$:

$$\frac{\epsilon}{\epsilon^*} = \frac{\lambda}{\beta} \left[1 - (1 - \beta\Gamma)^{1/\lambda} \right] \tag{14}$$

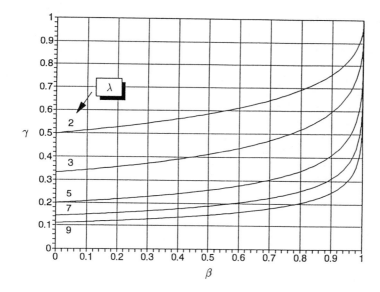

Figure 8: Variation of failure strain with stress level

and

$$\frac{\varepsilon_f}{\lambda\varepsilon^*} = \frac{1 - (1 - \beta)^{1/\lambda}}{\beta} = \gamma \tag{15}$$

where $\varepsilon^* = \dot{\varepsilon}_o t_f$ (the Monkman-Grant constant)

As for the results of the previous section on rupture times so here the Kachanov results are modified by the stress level σ_o at which the creep test is performed in relation to the stress $\bar{\sigma}$. These modifications are relatively small if $\sigma_o \leq \bar{\sigma}/2$ ($\beta \approx 1$) but they are significant above this level particularly insofar as the rupture strain ε_f is concerned. For high stresses $\gamma \to 1/\lambda$ which means that straining continues indefinitely until failure is precipitated by necking of the specimen. For intermediate stress levels, the value of the rupture strain can be grossly different from $\lambda\varepsilon^*$; typical values are shown in Figure 8. This could be significant in the "A-parameter" approach [14] which is reliant on the use of ε^* and ε_R defined by the unmodified results ($\beta = 1$). This is because the strain at failure depends on the initial stress level at which the test is performed in relation to the value $\bar{\sigma}$.

4.1.3 In-Plant Remaining Life Estimates

The ideal of assessing plant behaviour through the use of computers and laboratory data is unlikely to be accurate, appropriate or reliable, let alone sufficient for the operators responsible for maximising plant utilisation. Strangely enough, the most useful laboratory – the plant itself – is rarely used to collect pertinent data. Procedures for this are not planned for during the design stage, records of plant inspections during shutdowns are can be out-of-date and inappropriate. Maintenance technicians are not the right people to decide which measurements to make and where to make them or interpret them – they are excellent in their own techniques but that is all. Even the most rudimentary dimensions and measurements can provide information about current conditions and act as guidance for future behaviour. Pipe diameter growth, surface replicas, interferometric results are examples of the many simple techniques that can be used. As an illustration, consider how the results of the simple analyses given earlier could be used systematically in dimensional measurements conducted regularly over a period of time.

It is assumed that suitable proven measurement techniques have been used such that strains in a component are monitored with time. These values could be available from plant startup – an unlikely situation – or some time later. The object is to make estimates of the remaining design life. Schematics in Figure 9 typify these situations.

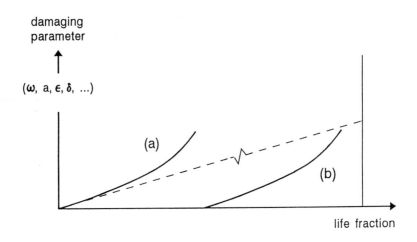

Figure 9: Schematic of damaging parameter ((a) from start-up or (b) after start-up)

From laboratory data and a knowledge of the operating conditions, values of $\bar{\sigma}$ (from short-term tensile testing), r and p and hence λ (equation 13) and β (equation 10) enable the following procedure:

   i. calculate $\dot{\varepsilon}_o = \dot{\varepsilon}_o(\sigma_o)$ and $\lambda = \dfrac{1+r}{1+r-p}$

  ii. measure $\varepsilon/\dot{\varepsilon}_o t$ at some time when $\varepsilon > \dot{\varepsilon}_o t$

 iii. calculate Γ from equation (14) rearranged as :

$$\frac{\varepsilon}{\dot{\varepsilon}_o t} = \lambda \left[\frac{1 - (1 - \beta\Gamma)^{1/\lambda}}{\beta\Gamma} \right]$$

and using the value of $\varepsilon/\dot{\varepsilon}_o t$ which has been measured

iv. calculate $t_f = t/\Gamma$

v. repeat steps iii and iv and take the average value of t_f

vi. check the current strain rate from historical data and check the measured value using equations (7) and (11)

vii. iterate items iii - vi and repeat the process over periods chosen as experience grows. Each time re-evaluate estimates of t_f ; these should converge

viii. extract "boat"-like samples, if possible, and check cavity densities with the results of equations (7) and (11)

Check results with ultrasonic or any other data and plot the progress of ω values remembering that its critical value is $1 - \sigma_o/\bar{\sigma}$ (from equation (4)).

Of course, the suggested procedure is over-simplified and the experimenter will generate his own systematic techniques to gain the most reliable results from consistency tests. In any event, the results obtained will be more useful than having no results at all. At the same time, other features of the plant behaviour will be discovered. The most important of these will be the appearance of cracks, most particularly towards the end of the design life and into regions of life extension. On the basis of best estimates of crack profile measurements, aids to remaining life assessments through the use of mechanics can be made of failure probabilities.

4.1.4 Probabilistic Structural Analysis

Most structural analysis methods are deterministic in their inputs and outputs – whether they are hand-book based or use computer programs. The key to performing effective probabilistic analyses of component response is not to generate even more complex and expensive analysis programs but rather to use existing deterministic methods combined with a systematic perturbation sequence in which post-processing gives the required probabilistic response. This has been successfully achieved in a recent doctoral thesis[15] where the necessary algorithms have been developed and automated. In addition, the same techniques can be used in a "manual" approach wherein the random parameters of the model are perturbed manually for each perturbed evaluation.

These exciting developments are being extended now in order to deal with typical problems, such as:

- design/remaining life assessment using the creep/damage analyses dealt with earlier in this paper

- safe inspection interval evaluation

- leak-before-break analysis

and in typical components which include:

- pressure containing systems, boilers, headers, heat exchangers

- rotating machinery

- transportation structures containing cracks ; aircraft, ship hulls, railway wheels

- large structures with random environmental loadings

The scheme has already been tried on the large scale general purpose finite element program ABAQUS and also a general purpose fracture mechanics code. These trials have served to prove that a combination of knowledge of mechanics and of statistical processes can be successfully and economically applied to any readily available structural analysis code.

Whilst the benefits of these methods are obvious, there are some powerful inhibitions to be overcome before they are widely used. One of these is the lack of communication between engineering practitioners and plant operators ; the latter of these are wary of statistical analysis whilst their managers often have little engineering knowledge. Another is that the owners of plant are only just beginning to realise that huge sums of money can be saved by life extension through good design for failure prevention. In both cases poor communication and out-dated methods of engineering education have to be overcome. Perhaps educators, owners and practitioners need to heed the words of a well-known engineer and educator:

"You cannot learn everything. The objects of knowledge have multiplied beyond the powers of the strongest mind to keep pace with them. You must choose between them and the only reasonable guide to choice in such matters is utility. What I deplore in our present higher education is the devotion of so much effort and so many precious years to subjects which have no practical bearing on life".

J A Froude,
St. Andrews, 1868

5 Concluding Remarks

There is little that is new in this paper. This feature was one of the objects of the paper and to reinforce the view that more pragmatic ways for designing against failure are needed. Most of the concepts for this are well known but the tendency for specialists to overlook these in favour of embroidery is often prevalent. There is clearly a need to engage more able people in the process of design so that the inextricable links between mechanics and materials are reinforced by generalists and not specialists. The wasteful procedures of using expensive, overly sophisticated laboratory equipment to gather data must be reduced in favour of in-field testing.

This whole procedure will require active cooperation between companies, plant operators and research workers in order to stem the strong flow of able people back to the cloisters where they are able to specialise without accountability. Froude's suggestion of utility as a criterion for pragmatic allocation of resources is probably more apt than it was when he made his suggestion. The warning of Froude has gone unheeded and so, it seems, has that made more recently by D C Drucker :

"The strong steady drift of far too large a fraction of the best graduates out of most branches of engineering into modern electrical engineering, out of electrical engineering into theoretical physics and then into mathematics is evidence of a contemporary form of medieval scholasticism".

The pressing problems involved in designing against failure and particularly those of the life assessment of ageing plant will increasingly affect all members of the community. This is now evident in Eastern Europe and is also particularly true in developing countries where resources in terms of able engineers and capital are reducing very quickly. The role of engineers who can make pragmatic contributions must be strengthened in preference to the self-evident profligacies of research activities in which wheels are too often being re-invented. International collaboration is vital.

6   Acknowledgement

The help of M A Weber during the preparation of this paper is acknowledged.

7   References

1. Robert Hooke. "De Potentia Restitutiva", London, 1678

2. August Wöhler. "Ueber Festigkeitsversuche mit Eisen und Stahl", Berlin, 1870

3. L Kachanov. "On the Time to Failure in Creep Rupture", Izv. Ak. Nauk SSSR Otdel, Tech Nauk 8, 1958

4. Y Rabotnov. "Creep Problems in Structural Members", North-Holland Publ. Co., London, 1969

5. R K Penny and D L Marriott. "Design for Creep", McGraw-Hill, London, 1971

6. "Assessment Procedures for High Temperature Response of Structures - R5", Nuclear Electric p.l.c., Berkeley, UK, 1992

7. K Nishida, K Nibkin, G Webster. "J Strain Analysis", vol 24, 1989

8. R K Penny and J Gryzagoridis. "Holographic NDE in Pressure Vessels and Piping", To be published by Int. J. Press. Vess. and Piping, Elsevier, London

9. J L Chaboche and J Le Maitre. "Mecanique des Materiaux Solides", Dunod, Paris, 1985

10. R G Sim. "Creep of Structures", PhD dissertation, University of Cambridge, UK, 1968

11. R K Penny and M A Weber. "Robust Methods of Life Assessment during Creep", Proc. CAPE '91 Colloquium on Ageing of Materials and Lifetime Assessment, Elsevier, London, 1992

12. A R S Ponter. "Derivation of Energy Theorems for Creep Constitutive Relations", Proc. IUTAM Gothenberg 1970, Springer, Berlin, 1972

13. A Wickens and G Oakes. "High Temperature Properties of Creep Resistant 12Cr Steam Turbine Blading Steels", Proc. I Mech E Conf. on Creep, Sheffield, Sept, 1980

14. B J Cane and P F Aplin. "A Mechanistic Approach to Remanent Creep Life Assessment", ASME J Press. Vess. Tech., vol 107, 1985

15. M A Weber. "Stochastic Structural Analysis of Engineering Components Using the Finite Element Method", PhD Thesis, University of Cape Town, 1993

3 EVER DIMINISHING CONCENTRIC CIRCLES: THE BUSINESS OUTLOOK FOR ENGINEERING CONSULTANTS IN THE LITIGATION AREA

BERNARD ROSS

Failure Analysis Associates, Menlo Park, California, USA

The worldwide economic compression of the 90s, which currently affects both industry and the professional services business, has also been experienced acutely in sectors of the legal and engineering communities that deal with product liability, technical insurance, and failure analysis matters.

The operating phrases here are downsizing, increased competition, enhanced productivity and a far more intense scrutiny of direct labor, overhead and ancillary costs to the client. Major law firms are confronting the previously unthinkable process of not only laying off associates but actually terminating partners. The ripple effect of these actions is being propagated throughout the litigation-related consulting arena. The retention process has changed dramatically and no longer do insurance companies, acting through their attorneys, engage experts on an a priori unilateral basis. A growing demand for fixed price job proposals which forecast overall project costs, previously open-ended, is now de rigueur. Multiple source solicitations and run off competitions are commonplace. Frequently the consultant is required to present project plans and expected findings in a pre-selection conference, without compensation, a notion that was non-existent in the 80s.

Another factor curtailing expert witness billings is a decided move in the United States toward avoiding elaborate court trials which require costly exhibits, video animations, laser disk productions, photo blowups and the like. Nowadays lawsuits which do not settle out of court are often litigated in the more professional, and as a corollary, less theatrical setting of judicial mediation before retired judges and moonlighting attorneys. Such procedures result in considerable loss of both consulting time and cost revenues to the independent professional. Rehearsal and preparation effort committed to the presentation of trial testimony is substantially reduced when the trier of fact is not the lay jury.

According to Henderson and Twerski in their seminal New York University Law Review paper *Closing The American Products Liability Frontier: The Rejection of Liability Without Defect*, product liability litigation has now reached its outermost frontier. To quote from their article:

Structural Failure: Technical, Legal and Insurance Aspects. Edited by H.P. Rossmanith. Published 1996 by E & FN Spon, 2–6 Boundary Row, London SE1 8HN, UK. ISBN: 0 419 20710 4.

For over one hundred years, American courts expanded the rights of plaintiffs in products liability cases. First the courts eliminated the privity requirement, next the necessity of proving fault, and finally, the necessity of proving a production defect. The next logical step in this progression would be to eliminate the need to show any type of defect at all. This step cannot and will not be taken. The authors describe how a system of liability without defect would work, and then they demonstrate why such a system is neither workable nor desirable. They also discuss the judicial system's flirtation with such an expansion and that judicial system will not tolerate this development in products liability law.

A major impetus for savings from the monetary standpoint has been the exorbitant losses suffered by the insurance industry, during past five years. Renowned disasters such as Hurricane Andrew (Florida), Hurricane Iniki (Hawaii), Exxon Valdez, the San Francisco Loma Prieta Earthquake, the Los Angeles Northridge Earthquake, and so on, have taken their toll on insurance reserves. Apropos, Lloyds, in a recently released accounting of 1991 financials, posted a record shortcoming of almost 3 billion pounds.

A further aggravation for insurers has been the excessive, even egregious expenses incurred by defense law firms as well as substantial outlays for engineering, economics, and medical experts when litigating property damage, personal injury, and wrongful death actions. The derisive term Skaddenomics was devised to define such largess using as paradigm conferences at the Skadden Arps Wall Street law firm whence a 50-cent cup of coffee fetched from an outside purveyor was charged to clients prorata at $85 per hour delivery time. Other complementary burdens including secretarial labor charges for copying, excessive fax and telephone unit costs, and the like were all part of this abscess.

The inevitable result of these outlandish billing practices has been an inexorable upsurge in cost consciousness by insurers with concomitant reduction for experts' services in an already crowded consultant marketplace. Whereas, in the past, experts were routinely retained ad hoc via initial call from a lawyer or insurance company, it is now commonplace for clients to contact many potential candidates whilst soliciting bids for their services so that final hiring is made on a competitive basis. Multi sourcing obviously acts to the disadvantage of qualified individuals who over prior years have built up a substantial experience base and professional reputation with corollary high hourly charges. By nature, professional advisory services are traditionally offered on a time and materials basis. The new regimen is one of fixed price, both incremental and global, that is, a firm cost ceiling must be agreed upon for the various phases of an accident investigation such as initial review of documents, site inspection, analysis, testing, travel, and so on.

Insurance companies and large corporate entities, particularly the automobile manufacturers, have embarked on a further cost cutting campaign whereby heavy reductions in the number of regional counsel allied through retainer with the parent company have taken place. For example, General Motors has decreased the number of law firms representing their legal interests across the U.S. approximately ten fold. On the other hand, Alcoa Corporation, in an entirely novel approach, has engaged a major Wall Street law firm to handle their entire litigation portfolio for a fixed fee of $6 million per annum. It follows that a diminished attorney base results in far fewer opportunities for both renewing old contacts and establishing new liaisons.

Established consultants face increased competition from other engineers who are entering the expert witness marketplace. In particular, the large scale layoffs engendered by defense spending cut backs in the U.S. has unleashed a considerable number of unemployed professionals into the consulting fray. Engineering societies such as ASME and AIAA now promote seminars at both regional and national conferences titled, How to Become an Expert Witness. Another source of emerging contention is the growing number of company engineers who, upon reaching retirement age, launch out on their own as forensic consultants and expert witnesses. Often these are the very same individuals with whom the outside expert has dealt on a professional basis during prior years while protecting the defense interests of the engineer's own corporate employer.

High technology has pervaded the workplace and attenuated consultants' billings. Depositions, now taken by court reporters with computer-assisted word processors, no longer have to be scanned in a labor intensive way to isolate specific passages of testimony. The search, retrieval and printout of such information is conducted in essentially zero time by low cost electronic equipment

About a decade ago the use of computer engineering graphics as an enhancement toward delivering highly technical testimony to a lay jury was both innovative and insanely expensive. For example, Failure Analysis Associates created a video animation of the World Airways DC-10 landing overrun into Boston Bay at a cost of almost $200,000. Today, that same tape can be produced by modern computer graphics technology for about one-fifth the cost, most economically through specialty firms that are solely in the computer visual aids business. Also commonplace now, laser disc light pen activated graphics which is becoming more prevalent as a means of communication between the culture of engineering and the legal, insurance and lay audiences.

The ongoing process of change in tort law itself has curbed the expansion of plaintiff's liability theories toward pursuing lawsuits against manufacturers and insurers. In fact, the product liability field is seemingly in process of a restructuring whereby limited recovery rights, more liberal allocation of comparative fault, bars on subcontractor employee claims against general contractors and property owners, and a general tenor of political conservatism, have all acted to diminish opportunities for engineering consulting retentions on accident cases.

Not withstanding all the foregoing, 1993 witnessed some of the largest plaintiff's verdicts ever against corporate and insurance defendants. These decisions included $1.2 billion won by Litton Systems claiming Honeywell infringed on its patent for the ring laser gyroscope as well as $105 million and $100 million verdicts in two separate vehicle fire crashes, the former in Atlanta, relating to GMC pickup gas tank placement and the latter in New York City the result of a police car chase. On the environmental front, plaintiffs were awarded $45 million against Combustion Engineering for toxic residue contamination of rural farm and timber land from a kyanite mine. In perhaps a case with comical overtones, plaintiffs were successful in gaining a $75 million jury award against Domino's Pizza for an accident caused when their delivery truck speeded through a red light in order to meet the company's "Within 30 minutes or less delivery guarantee."

New areas of technological importance are also emanating as part of the litigation process. For example, environmental lawsuits embracing ground contamination, toxic waste, and both air and noise pollution are fast growing areas which require different technical expertise than heretofore related to classic product liability cases. The very controversial epidemiological problem of electro magnetic radiation effects on persons in homes under high voltage transmission lines is an expanding realm of near future litigation practice.

Of course, as typical in engineering, science, and technology, when old disciplines fade, new challenges evolve and are substituted. The coming field for professional consultants, undergoing rapid generic development at this time, is risk auditing, risk analysis and disaster management supported by an insurance industry beset with enormous portfolio losses due to poorly written policies that have not gained benefit from quantitative assessment of risk. This dislocation has been particularly grave in earthquake risk coverage where a paucity of integrated knowledge concerning the response of buildings and other structures to seismic loadings has not been coordinated with the simplistic seismic zone intensity maps favored by the underwriters.

Consultants with structural, seismic, and statistical skills are now involved in developing software systems that can analyze earthquake, hurricane, tornado and other natural disaster risks, both on the basis of probability and using individual catastrophic occurrences which incorporate tne chances of different magnitude events developing on diverse faults or fronts within a specified future time period. Since these analyses coordinate both actuarial and scientific methodologies, liaisons between the engineering, legal and technical insurance communities are being strengthened in a synergistic way so as to provide economic and professional advantages for all three disciplines.

As a final note, it is the author's very subjective opinion that the insurance and litigation defense industry is affected by an enigmatic case of penny-wise, pound foolish astigmatism. Whereas, insurance claims agents gain satisfaction from holding costs for outside experts to a minimum, they seem oblivious or indifferent to the excruciating losses which result in the courtroom from inadequate research, analysis, testing, and trial preparation. The common agenda is to blame lawyers and their experts for the courtroom defeat whilst the incongruous penny pinching benefits of controlling minimal defense costs accrue to the claims manager.

4 PRODUCT LIABILITY AND THE EXPERT WITNESS: A CANADIAN PERSPECTIVE

IAIN LE MAY

Metallurgical Consulting Services Ltd, Saskatoon, Canada

Abstract

The Canadian practice in the use of expert witnesses in product liability cases is described, together with a brief review of the Canadian legal system. It is emphasized that the bringing in of an expert at an early stage can lead to reduced costs and out-of-court settlements in many cases.

1 Introduction

In product liability cases in Canada the use of expert witnesses is normal: in this respect the procedure is similar to that followed in the United States. There are, however, many differences between the two countries.

First, the damages awarded by Canadian courts are much less than in the US, thus the pressure for legal action is reduced greatly. Second, contingency cases are not normal, and are discouraged in most of the Canadian provinces, so there are far fewer "ambulance chasers" among the lawyers. Also, we rarely see class action lawsuits taken out on behalf of a group of people who may have been peripherally affected by an accident: most of these in the US are initiated by lawyers who may advertise for clients affected by the accident.

A product liability suit will normally be held before a judge alone in Canada. This reduces the length of the trial and reduces costs as compared to a jury trial. It also means that there is less posturing and playing to the gallery, i.e., the jury, on behalf of lawyers and expert witnesses alike. The testimony can usually be presented in a more detailed manner and judges who have been appointed from the ranks of senior barristers rather than elected, are usually much more able to make a sound and rational judgement on the basis of scientific evidence than can the average jury.

The system also differs from the US with respect to the nature of the discovery process. In most jurisdictions in Canada the various parties involved are represented by one responsible individual who is examined by opposing counsel, prior to the trial. The discovery material, and undertakings to provide additional information arising from the discovery, provide a great deal of information relating to the case before a trial commences, and in a pre-trial conference, presided over by a judge (not the one who

Structural Failure: Technical, Legal and Insurance Aspects. Edited by H.P. Rossmanith. Published 1996 by E & FN Spon, 2–6 Boundary Row, London SE1 8HN, UK. ISBN: 0 419 20710 4.

will preside at the trial itself), an attempt will be made to have the parties settle rather than spend (or waste) time in undertaking additional preparation and then going to court for the trial itself. The expert witness or witnesses will not be required to be examined in a deposition prior to trial, as in the US, but their opinions are normally required sometime before the trial, and normally before the pre-trial hearing, to enable all parties to evaluate their chances of winning the case or seeking a negotiated settlement.

As a result of these factors, experts in Canada may expect to spend considerable time in preparation for trials and very much less in court. The number of court appearances is probably much less than would occur in the US.

2   The Canadian legal system

As in the United Kingdom and the United States, there are two sources of law. These are common law and statute law. The former is based on the British system and on rules and precedents from the courts, and is therefore formed by the judiciary. The latter is formed from legislative actions of federal, provincial and municipal bodies. In the Province of Quebec, The Civil Code is based on the French system and statute law and is much more comprehensive than the statute law in other provinces.

The highest legal body in Canada is The Supreme Court, followed by the Courts of Appeal and Trial Courts of each province. In addition, there is a Federal Court which deals with matters under federal jurisdiction: there is a Federal Court of Appeal and appeal from it also is to the Supreme Court. Each province also has provincial courts that deal with criminal law, family law, young offenders, provincial statute violations and small claims, these being subject to appeal and overruling from the federally appointed courts of each province.

While the Supreme Court decisions are binding on all courts, the decisions of one court are persuasive only on another court of the same level in another province. Nevertheless, precedents established in one jurisdiction or province are generally applicable in another, or at least may be used and argued as a precedent in the other court.

Precedents established in courts in the United Kingdom and the United States are permissible in Canadian courts, so that it is not just Canadian precedents that count, although they are considered more persuasive and more applicable.

The expert witness will normally find himself or herself giving evidence in a Provincial Trial Court, from which any appeal will be made to the provincial Court of Appeal. On occasion, the expert may be called upon to testify in a small claims court, but this is rare as costs are low in these and the level of claim can only be of a few thousand dollars. It is not normal that an expert will be involved in an Appeal Court hearing. Appeals are made based on points in law, for example that a judge has erred in dismissing or including some evidence or in a ruling in the lower court. Only if there is some specific new evidence to be presented that has come to light subsequently, might the expert be involved.

3   The role of the expert witness

As noted previously, the major role of the (potential) expert witness in a product liability case is pre-court. The earlier the stage at which the expert is involved the better, but all too often the involvement is commenced only when the insurance companies or lawyers concerned in a case cannot reach agreement, and the failed part has been handled many times. Indeed, the expert may be contacted to determine how a specific part broke without all the evidence being available as it may have been disposed of.

When a product failure is investigated it is important that the expert investigator, who is most often an engineer with a background in mechanical engineering, materials and metallurgy, or both, approaches the investigation with the thought that the case may end up in court. By the keeping of proper records, careful documentation including sketches and photographs, and thorough and complete investigation, the preparation of a case is facilitated greatly. If the case is not a good one the lawyer concerned must be told immediately, so that a settlement can be sought. Similarly, if the case is a strong one, the technical evidence complete and well prepared, then a settlement can be sought on a favourable basis prior to the case going to trial.

Trials are expensive and it should be the objective of the expert to aid his client to obtain the most favourable settlement consistent with the scientific evidence that can be offered.

When conducting an investigation of the failure analysis type, the expert's conclusions should be the same independent of which party is being represented. However, in preparing for trial, the expert will be involved in suggesting criticism of the approach of the expert or experts for the opposing party, and in suggesting questions for cross-examination.

In court, when a case does go to trial, the expert witness enjoys a privileged role, as opinion evidence can be presented. Generally the expert will be treated with respect by the judge unless he or she tries to be less than forthcoming with evidence and opinion. Such actions rapidly destroy the "expert's" credibility and the possibility of future expert investigation and testimony. In a similar vein, if the expert has made a less than complete study, this will be pounced on by the opposing lawyer and may also lead to comments by the judge. If the investigation is not done completely enough, the expert must tell his counsel beforehand and indicate an unwillingness to act as a witness unless and until the investigation has been completed properly. (Sometimes an incomplete investigation results from financial constraints imposed by the client but, if that is the case, neither lawyer nor expert should indicate a willingness to continue with the case.)

The writer has been in the position when it has been suggested by opposing counsel that more could have been done, and more detailed investigation made, but has countered this successfully by arguing that enough has been done, the facts and evidence are clear, and that further work would only have been a waste of the client's money without any additional evidence of value being produced.

On occasion, an expert who has conducted part of an investigation but is not being called as a witness, may be subpoenaed by the other side. This creates a difficult situation as the work done for a lawyer in anticipation of litigation is considered privileged and not subject to disclosure to the opposing party. However, the court may require the expert to testify, and files may be required to be disclosed through the issue of a Subpoena Duces Tecum, requiring all relevant documents and files to be brought to court. Quite apart from the difficulties all of this may create for the expert concerned, there is no guarantee of payment for the time involved, and only conduct money for attendance in court is <u>required</u> to be paid. This appears to be an area in need of clarification and resolution at the present time.

In an interesting Canadian case that will not be identified specifically, because there is a related case still before the courts, an expert witness for the defence has been sued for conspiracy to defraud the plaintiff. Action has also been commenced against the insurance adjusters and others involved in denying the plaintiff's claim. The judge in the case disallowed the expert's evidence on the grounds that he was being sued and that there was a potential for bias, notwithstanding the fact that the opinion evidence itself appeared to be sound. This interesting precedent suggests that an expert witness's testimony can be discounted by the party on the opposite side starting an action against the potential expert witness, and it is to be hoped that this ruling will not be used as a precedent or will be struck down.

4   Concluding remarks

In concluding these remarks, I would say that the expert witness can and does provide an important service to the Canadian courts in product liability cases. Currently, it often takes a long time for a case to go to court (in some exceptional cases up to ten years, where matters are complex) and court costs are high. With greater use of experts at an earlier stage and a movement towards alternate settlement procedures such as arbitration, costs and time delay can be reduced. With expert opinion available at an early stage, it is to be hoped that more out-of-court early settlements will result, and this should be of benefit to all concerned.

5 THE STATUS OF PRODUCT LIABILITY IN THE USA

J. SCOTT KIRKWOOD
Wingate College, Wingate, North Carolina, USA

It is too soon for studies to be conclusive, but indications are that the number of product liability suits instituted is declining. This conclusion is based upon observations over a number of years and having followed journals, court reporter series, and periodicals. Even more significantly, it appears that certain norms are being established, by the judicial system, economists, and by society in general. No longer is every case a new, precedent setting tort. The pendulum has swung from extreme to extreme, but now seems to be settling into a defined arc. However, there are still a number of points and questions which merit our interest.

Share of the Market Theory

The "share of the market" theory allows a plaintiff to hold each of a number of manufacturers/sellers liable for harm, in proportion to the share of the relevant market each defendant enjoyed at the time the harm occurred. The harmful product can be identified, but the specific manufacturer/seller cannot. This rule gained notoriety in the 1980 case involving the use of the drug diethylstilbestrol (DES), ingested by pregnant women to prevent miscarriages.[1] However, the harm appeared only in female children of those who had taken the drug, and after the children attained puberty, a generation later.

There is no federal rule or statute regarding nation-wide application of "share of the market" theory. Therefore, its use varies among the various states of the United States. It has been adopted by Calfiornia,[2] Florida,[3] New York,[4] Washington,[5] and Wisconson.[6] It has been rejected by the states of Illinois,[7] Iowa,[8] and Missouri.[9] A number of states have not yet addressed the issue. The concept holds a defendant liable when there is no proof that he specifically created the harm; this is a missing link in the chain of causality. The courts seem to recognize a social problem which demands an answer. "Market share liability" alleviates the problem, but does not provide an acceptable answer.[10] There is general reluctance to extend this concept to other commodities,[11] and there is strong concern that courts are improperly developing public policy which should be left to the legislature.[12] Judge Jack B. Weinstein, U.S. District Court for the Eastern District of New York, recently held that a DES manufacturer was subject to suit in New York state, although that manufacturer's product had never been sold in New York state. With reference to the federal Commerce Clause and the laws of New York, he asserted: "A substantial interjection of products at

Structural Failure: Technical, Legal and Insurance Aspects. Edited by H.P. Rossmanith. Published 1996 by E & FN Spon, 2–6 Boundary Row, London SE1 8HN, UK. ISBN: 0 419 20710 4.

any point of the national market has ripple effects in all parts of the market."[13]

The General Motors (GM) Pickup Recall

On Friday, April 9th of this year, the United States National Highway Traffic Safety Administration (NHTSA) requested the General Motors Corporation to recall an estimated 4.7 million pickups (i.e., light trucks) on which the fuel tanks were located outside the structural frame. This is one of the largest number of vehicles in any recall in history. The pickups were manufactured by GM between 1973 and 1987. Estimates of cost of relocating the fuel tanks are as much as $1 billion.

It is alleged by the government that the "side-saddle" location of the fuel tank makes the pickup extemely vulnerable to fire and/or explosion in the event of collision. After 1987, GM placed the tank inside the structural frame. There have been a number of suits for wrongful death or serious injury with multi-million dollar awards, and 36 large class action suits are now pending. To date, GM has refused to recall any of the trucks, contending that "The trucks meet all federal safety requirements."

On July 19th, General Motors announced a novel and innovative proposal to settle the class actions: It would give a $1,000 certificate good toward purchase of a a new GM light truck to every owner of its 1973-1987 full-size pickups. This, of course, does not settle the question of recall as sought by the government, nor does it get the allegedly dangerous trucks off the highways. Bill O'Neill, a GM spokesman stated: "What this basically does is separate the customer satisfaction issue from the technical questions." Of course, the certificate has value only when applied to the purchase of a a new truck. One plaintiff's attorney who opposes the plan suggests: "Why not give the truck owners money and the give the lawyers the coupons?" Cost of the certificate proposal is estimated at $6 billion; however, it is indeed a unique application of a sales promotion rebate! It is estimated that plaintiff's attorneys fees of $21 million will also be paid. The settlement is subject to approval by the U.S. District Court in Philadelphia. Hearings on the proposal had not been held as of this writing.[14]

Liability of Publishers for Physical Harm

Most suits are against the publisher, rather than the author, of books on the theory of "deep pockets" (since a judgment is worthless if against one with no assets). Courts generally hold that erroneous information from a a book or other publication is not a "product;" therefore, there is no "product liability;"[15] to hold the publisher liable would have a discouraging "chilling effect" upon the First Amendment rights of free speech and freedom of the press.[16] However, in a recent case the publisher of Soldier of Fortune magazine was held liable for publishing an ad placed by a mercenary killer which resulted in the murder by gunshot of one Richard Braun.[17]   The trial court instructed the jury that the defendant magazine could be found liable in negligence if the ad ..."presented a clear and present danger of causing serious harm to the public from violent criminal activity." The award of $4.3 million was upheld by the Court of Appeals Fifth Circuit; the U.S. Supreme declined to review.   A reduced amount of damages has since been negotiated by the plaintiffs (sons of the victim).[18]

The "Jeppesen cases" are a group of actions against the publisher of aeronautical navigation maps, which are a graphic presentation of tabulations prepared by the U.S. government.   The courts have held such maps to be "products,"[19] but have declined to provide a "clear light" to identify situations of liability.[20]

PUNITIVE DAMAGES

Punitive damages are a penalty for a <u>past</u> wrong and a deterrent to others who may be tempted to commit a similar wrong in the <u>future</u>.   The U.S. Supreme Court has ruled that punitive damages are not an excessive fine,[21] has declined to draw a "mathematical bright line" as the limit of reasonableness,[22] and in the recent TXO[23] case held that punitive damages of 526 times the amount of compensatory damages was NOT <u>per se</u> unreasonable and did not violate the due process requirement of the U.S. Constitution.

The TXO case was an instance of fraud and bad faith involving interests in land.   Compensatory damages of $19,000 were awarded, and punitive damages of $10 million, a ratio of 526!   The awards were consistently upheld through the U.S. Supreme Court.   If limitations were to be imposed, they must originate with the legislatures, not the courts.

Some 29 of the 50 states have imposed some type of statutory limitation on punitive damages.[24]   Some states have enacted statutes giving some portion of punitive damages to the state, although constitutionality of such is often questioned.[25]   The state of New York reports that its 20% share of awards of punitive damages has generated far less than the anticipated $10 million per year.[26]

FEDERAL APPROVAL AS A SHIELD FROM PRODUCT LIABILITY

This concept seems very similar to the "state of the art" defense, except the government agency establishes and specifically defines the "state of the art."

There have been a series of recent cases alleging injury from collagen which, which injected into the lip, creates the alluring "pouting" appearance.   The First[27] and Fifth[28] U.S. Circuit Courts of Appeal have ruled that product liability claims involving Class III medical devices as defined and covered by the 1976 Medical Device Amendments (MDA) to the federal Food, Drug & Cosmetic Act are preempted.[29]   The Act reads, in part: "No state or political subdivision of a state may establish a safety or effectiveness requirement for a device subject to the provisions of the MDA which is different from or in addition to any requirement under the MDA." Since 1976, these devices subject to the Act have been required to undergo extensive testing before being placed in the market. Preemption was held to deny recovery not only on the theory of "failure to warn," but also on breach of warranty, negligence, misrepresentation by the First Circuit. In holding the preemption, the court stated that "Any state requirement which, in effect, established a new substantive requirement for the device in a regulated area such as labeling, is preempted.[30] The Fifth Circuit found preemption of claims of defective manufacture as well as labeling and failure to warn.

Somewhat similarly, the Federal Insecticide, Fungicide and Rodenticide Act,[31] (FIFRA) has been held to preempt state common law actions on inadequate labeling.[32] This increase in the preemption by federal law of state "failure to warn" actions appears attributable to the landmark <u>Cipollone</u>[33] case which turned on the application of the Federal Cigarette Labeling and Advertising Act.[34] In <u>Cipollone</u> the U.S. Supreme Court said that when the U.S. Congress enacts a statute containing an express preemption, the courts should assume that no other preemptions are intended. Further, there is a general presumption against preemption. Thus, FIFRA has been interpreted to not preclude suits in causes <u>other</u> <u>than</u> labeling, such as defective design or advertising.[35]

Yet to be resolved are literally thousands of cases alleging billions of dollars in damages because of "failure to warn" of the consequences of silicone breast implants. These are <u>not</u> preempted because they were on the market well before 1976, when the Medical Device Amendment became effective. In fact, it was only in 1988 that the federal Food and Drug Administration ruled that such implants were class III devices and therefore subject to the strict tests and regulation under the Act.[36]

THE TOBACCO CASES

The battle over injury from the use of tobacco still rages. Tobacco is a significant agricultural crop in the southeastern United States. A subsidy is still paid to farmers who grow it. In its 1993 World Development Report the World Bank attributes 3 million premature deaths a year to the use of tobacco. Consumption is <u>decreasing</u> in industrial countries while <u>increasing</u> in developing countries. In the United States, advertising of tobacco products on TV or radio is prohibited. Cigarettes will kill an estimated 435,000 U.S. citizens each year from cancer, emphysema, and heart disease, and will burden the nation with an added health care cost of 65 billion dollars![37] In his proposed national health plan, President Clinton would increase the federal tax on a pack of cigarettes from 24 cents to 99 cents, four-fold plus! The Associated Press on October 26th announced that the Tobacco Institute, the Washington,DC trade group for America's cigarette makers, is sharply reducing its staff from 83 to about 50 and closing a tentative eight of eleven regional offices.[38]   Legislation to ban smoking in all public facilities...that is, private as well as government owned spaces regularly entered by 10 or more people at least once a week... unless there is a designated smoking area with a separate ventilation system, has been introduced in each of the houses of the U.S. Congress.[39]   There have been many cases tried, many before a jury, and pursued through appeal.   Despite this, the tobacco interests have prevailed.   Not one plaintiff has recovered <u>any</u> amount of damages for injuries as a result of the use of tobacco.

A FEDERAL PRODUCT LIABILITY LAW???

There are a number of significant studies and endeavors now in progress in the field of product liability.   Topics include the alleged "explosion" of punitive damages,[40] the <u>fairness</u> of the massive class actions in asbestos litigation,[41] the drafting of the American Law Institute's (ALI's) <u>Restatement (Third) Torts: Products Liability</u>,"[42] and the development, efficacy, and effect of product warnings.[43].   Of all the studies and proposals, most would agree that by far the one with greatest effect and with greatest impact upon the economy of the United States would be a <u>federal, uniform</u> product liability law.

For those not familiar with the United States of America, please permit a brief explanation.   The USA is composed of 50 states, its territories, and the District of Columbia.   It is indeed a union of states---a federation. The term "federal" is used to indicate the national government, rather than the states.   By definition in the written Constitution of the USA, the national or federal law is "the supreme law of the land," although this same Constitution <u>limits</u> the functional areas of authority of the federal government which were delegated to the federal government by the states.   There is no argument that the federal law is supreme in those areas delegated to the federal government by the states.   The U.S. Constitution also contains authority for the U.S. Congress to enact a statute containing a federal product liability law, which would bind all the states (and territories, and the District of Columbia, of course).   This is within the "Commerce Clause," among the <u>enumerated</u> powers of the Congress.

Thus, it is a <u>political</u> question as to whether the United States will enact a federal product liability law.   Last September, the Senate failed by two votes to move the bill from committee to the floor of the Senate, where a vote could be taken on it.   Thus, it "died in committee."   Senator Kasten, of Wisconsin, perhaps the bill's most ardent supporter, was defeated in his bid for reelection.   The Republican administration of President Bush was defeated by now President Clinton, a Democrat.   This was viewed by many as lessening the prospects of enacting any kind of federal product liability law.

However, a new Congress convened in January of this year and similar but not identical bills have been introduced in both the Senate and the House (of Representatives), in each instance with bipartisan support. Senator John D. Rockefeller, Democrat from West Virginia, introduced Senate Bill 667, which has since been referred to the Senate Committee on Commerce, Science and Transportation.[44]

The Product Liability Fairness Act, Senate Bill 667 (S.667) provides, in essence:[45]

a. Expeditious settlement, out of court and/or by Alternative Dispute Resolution (ADR) is encouraged, with some additional penalties if a party who declines such method fares less well at trial, except no penalty may be assessed against a plaintiff who declines ADR.

b. A seller is liable primarily for only his own negligence, or breach of an express warranty.

c. Punitive damages require proof by "clear and convincing evidence," that the harm was caused by defendant's "conscious, flagrant indifference to the safety of those persons who might be harmed by the product." Separate trials may be had for punitive and compensatory damages.

d. Pre-market approval or certification by the Food & Drug Administration or Federal Aviation Administration is a defense against punitive damages except in cases of bribery or withholding of relevant information. (This seems to be the government defining the "state of the art.")

e. Alcohol or illegal drugs are an absolute defense if the plaintiff's condition is such that he was more than 50 percent responsible for his injuries (i.e., contributory negligence).

f. Joint and several liability is abolished with respect to non-economic damages, such as pain and suffering.

g. The statute of limitations is two years, commencing when the claimant should have discovered the harm and cause. The statute of repose for capital goods is 25 years.

h. An employer's right to recover workers' compensation benefits from a manufacturer whose product allegedly harmed a worker is preserved, unless the manufacturer can prove that the employer caused the injury.

In the House of Representatives (the "lower house"), The Fairness in Product Liability Act of 1993 (House of Representatives Bill [H.R.] No. 1910) was introduced by Representative J. Roy Rowland, Democrat, of Georgia. Major provisions of the Bill are summarized:[46]

a. The statute of limitations starts to run when the harm or its cause is discovered, or should have been, and runs for 10 years. A statute of repose of 25 years is applicable to capital goods, <u>provided</u> the claimant is then eligible to receive workers' compensation for the harm.

b.   Sellers may be liable to injured consumers in negligence in case of alteration, assembly, or false representations.   They may be liable also if the seller has dealt with a foreign manufacturer that has no assets in the U.S. or has gone out of business.

c.   For punitive damages, a claimant must prove by "clear and convincing evidence," (a high standard of proof) that the defendants's conduct manifested a "conscious, flagrant indifference" to the public safety.   Punitive damages also may be litigated separately from liability and compensatory damages.

It is a defense against <u>punitive</u> damages for drugs and medical devices to have obtained pre-market approval of the federal Food and Drug Administration (FDA), in the absence of fraud, withholdinbg information, or bribery of FDA officials.

It should be noted that only Great Britain (not using a jury system on this issue) and the US allow punitive damages. Other nations generally do not allow them.

d.   H.R. 1910 would exempt NON-economic issues (e.g., pain and suffering, emotional distress), from the "deep pocket rule" (whereby any defendant liable for <u>any</u> portion of the harm may held liable for the entire damages).   The Bill would adopt the "California rule" and hold each defendant liable only in proportion to his share of the responsiblity for the harm.   This does NOT affect the joint and several liability for <u>economic</u> harm, such as medical costs and lost wages.

e.   There is no recovery for a plaintiff when a jury has determined that the plaintiff was the <u>primary</u> cause (i.e., more than 50%) liable for the injury.[47]

f.   A claimant's damage award will be reduced by a percentage attributable to misuse or alteration, as determined by a jury, pursuant to an instruction of the trial judge.

Thus, a user is to some degree held responsible for his own actions.[48]

g.   There is an inducement to the employer to provide a safe workplace.   Currently, an employer may recover workers' compensation he has paid to his employee, when the employee was injured by the product of a manufacturer (e.g., machine tools). This is done by the employer's filing a subrogation lien against the manufacturer.   This Bill would deny this recovery by the employer <u>when</u> the employer is at fault (e.g., inadequate training, removing a safety guard).

It is to be noted that the two bills are not identical. For example, the Senate bill includes alternative dispute resolution

(ADR) and expedited settlements; a defendant who rejects a plaintiff's offer to settle, then is directed by verdict to pay more, will also have to pay the plaintiff's fees and costs. The House bill would reduce the plaintiff's recovery if he has altered or misused the product.[49]

Whether a uniform (national) product liability law for the United States will be enacted this year is highly problematical---or, as we say, is "anybody's guess." From the perspective of a manufacturer or seller, the advantages of a uniform rule are well known and obvious. Yet there are a number of causes for concern and opposition to it.

Plaintiff's lawyers groups have been among the most vocal in opposing the legislation. The Rand Institute for Civil Justice, a California "think tank," published a study on September 23rd. It reported that although there were advantages to such legislation, ..."for most products, the impact of liability has had negligible effects on pricing."[50] There is doubt about the extent of the alleged "chilling effect" on new product development. Professor Lucinda Finley, of State University of New York at Buffalo School of Law, opines that the act ..."will not stop frivolous suits, 'runaway jury verdicts,' or product liability suits in general."[51] Professor Michael Saks of the University of Iowa School of Law notes that <u>federal</u> cases mostly have been grouped around a few items, such as asbestos, a few industries, and have actually declined in number.[52]

There is also now the general distrust of government and the concern for potential loss of sovereignty by the individual states. The European Community has witnessed this concern with regard to the Maastricht Treaty. The United States has had a deep concern over preservation of "states' rights" from the days even preceding the birth of the nation. In fact, it delayed the adoption of the Constitution. State Representative Mike Box of the state of Alabama said there is "no compelling reason to disrupt state law" and that a federal act "opens the door to ever more significant intrusions into state law."[53]

In a private conversation with a member of the United States Congress, the writer was informed that "If Congress is ever going to pass such a bill, <u>this</u> is the year to do it!" From personal observations and research, the conclusion is drawn that there simply is a <u>lessening</u> of discussion and dissatisfaction expressed about the problem of product liability. Or, perhaps, the tort system as a part of our law has actually served its purpose and further restrictions upon it are not needed. Unfortunately, as is common with most social legislation as well as new rules in the law, such development does not occur without some innocent persons having suffered. Further, concerns in the United States over such as foreign affairs, dispatch of the armed forces, and the proposed national health plan have taken precedence in the eye of the populace and of Congress. Will the United States Congress find the time to debate and pass a uniform product liability law this year? Maybe...MAYBE!

ENDNOTES

1. <u>Sindell</u> <u>v.</u> <u>Abbott</u> <u>Laboratories</u>, 26 Cal.3d 588, 607 P.2d 294
   163 Cal.Rptr 132, <u>cert.</u> <u>denied</u>, 449 U.S. 912 (1980)
2. <u>Id</u>.
3. <u>Conley</u> <u>v.</u> <u>Boyle</u> <u>Drug</u> <u>Co.</u>, 570 So.2d 275 (Fla. 1990)
4. <u>Hymowitz</u> <u>v.</u> <u>Eli</u> <u>Lilly</u> <u>&</u> <u>Co.</u>, 73 N.Y.2d 487, 539 N.E.2d 1069,
   549 N.Y.S.2d 941, <u>cert.</u> <u>denied</u>, 110 S.Ct. 350 (1989).
5. <u>Martin</u> <u>v.</u> <u>Abbott</u> <u>Laboratories</u>, 102 Wash.2d 581, 689 P.2d
   368 (1984).
6. <u>Collins</u> <u>v.</u> <u>Eli</u> <u>Lilly</u> <u>&</u> <u>Co.</u>, 110 Wisc.2d 166, 343 N.W.2d 37,
   <u>cert.</u> <u>denied</u>, 469 U.S. 826 (1984)
7. <u>Smith</u> <u>v.</u> <u>Eil</u> <u>Lilly</u> <u>&</u> <u>Co.</u>, 137 Ill.2d 222, 560 N.E.2d 324,
   (1990).
8 <u>Mulcahy</u> <u>v.</u> <u>Eli</u> <u>Lilly</u> <u>&</u> <u>Co.</u>, 386 N.W.2d 67 (Iowa 1986).
9. <u>Zaft</u> <u>v.</u> <u>Eli</u> <u>Lilly</u> <u>&</u> <u>Co.</u>, 676 S.W.2d 241 (Mo. 1984).
10. <u>Hymowitz</u>, supra n. (4).
11. <u>Santiago</u> <u>v.</u> <u>Sherwin-Williams</u> <u>Co.</u>, 794 F.Supp. 29
    (D MA, July 1992)
12. <u>See</u> an excellent summary of this concern in "Market Share
    Liability Theory" by Stanley Fuchs, Professor of Business
    Law, Fordham (NY) University, presented at the annual
    meeting of the Academy of Legal Studies in Business,
    Colorado Springs, Colorado, August 1993.
13. <u>Ashley</u> <u>v.</u> <u>Abbott</u> <u>Laboratories</u>, 789 F.Supp 552, (ED NY, 1992)
14. <u>The</u> <u>Wall</u> <u>Street</u> <u>Journal</u>, Apr. 12, 1993, p.A3; June 2, 1993,
    p.A3; June 23, 1993, p.A3; July 14, 1993, p.A6; July 20,
    1993, p.A3; Aug. 29, 1993, p.A2; Sept. 15, 1993, p.B8. Also
    see <u>Charlotte</u> <u>(NC)</u> <u>Observer</u>, July 20, 1993, p.2D
15. <u>Winter</u> <u>v.</u> <u>G.</u> <u>P.</u> <u>Putnam's</u> <u>Sons.</u>, 938 F.2d 1033 (CA 9th, 1991)
    (The failure to clearly identify poisonous mushrooms among
    edible ones.)
16. <u>Id</u>.   Also <u>Watts</u> <u>v.</u> <u>Bauer</u>, 109 Misc.2d 189, 439 NY2d 821
    (Sup Ct 1981).
17. <u>Braun</u> <u>v.</u> <u>Soldier</u> <u>of</u> <u>Fortune</u> <u>Magazine</u>, 968 F.2d 1110, (CA
    11th, 1992), <u>cert.</u> <u>denied</u> _____U.S._____, 113 S.Ct. 1028,
    122 L.Ed.2d 173 (1993).
18. <u>The</u> <u>Wall</u> <u>Street</u> <u>Journal</u>, March 1, 1993, p.B8.
19. <u>Saloomey</u> <u>v.</u> <u>Jeppesen</u> <u>&.</u> <u>Co.</u>, 707 F.2d 671 (CA 2nd 1983).
20. <u>See</u> an extensive review of this topic in "Publisher
    Liability for Erroneous Information Leading to Physical
    Injury" by William V. Vetter, JD, LLM, Assistant Professor,
    Wayne State University, presented at annual meeting of
    Academy of Legal Studies in Business, Colorado Springs,
    Colorado, August 1993.
21. <u>Browning-Ferris</u> <u>Industries</u> <u>v.</u> <u>Kelco</u> <u>Disposal,</u> <u>Inc.</u>, 492 U.S.
    257 (1989).
22. <u>Pacific</u> <u>Mutual</u> <u>Life</u> <u>Insurance</u> <u>Co.</u> <u>v.</u> <u>Haslip</u>, 499 U.S. 1
    (1991).
23. <u>TXO</u> <u>Production</u> <u>Corp.</u> <u>v.</u> <u>Alliance</u> <u>Resources</u> <u>Corp.</u>, 113 S.Ct
    2711, 125 L.Ed2d 366, 509 U.S. _____ (1993)
24. <u>See</u> "State Punitive Damages Statutes: A Proposed
    Alternative," Sandra N. Hurd, Associate Professor, and
    Frances E. Zollers, Professor, Syracuse University.
    Presented at annual meeting of Academy of Legal Studies in

Business, Colorado Springs, Colorado in August 1993.

25. Gordon v. Florida, 608 So 2d 800 (FL), 133 S.Ct 1647, 123 L.Ed 2d 268, _____U.S._____ (1993)cert. denied 3-22-93.

26. The Wall Street Journal, Nov. 13, 1992, p.B10.

27. King v. Collagen Corp., 983 F.2d 1130 (CA 1, 1993), cert. denied 10-04-93.

28. Stamps v. Collagen Corp., 984 F.2d 1416 (CA 5, 1993), cert. denied 10-04-93.

29. 21 USCA 360(c).

30. King v. Collagen, supra n. 27.

31. 7 USCA 136 et seq.

32. Papas v. Upjohn Co., 985 F.2d 516 (CA 11, 1993), cert. denied 10-04-93.

33. Lawyers Alert, Sept. 14, 1992, p. 24.

34. 15 USCA 1331 et seq.

35. Burke v. Dow Chemical Co., 797 FedSupp 1128 (ED NY, 1992)

36. Lawyers Alert, Sept. 14, 1992, p.24, (Vol 12, No. 18)

37. Charlotte (NC) Observer, Oct. 21, 1990, p.1A.

38. Charlotte (NC) Observer, October 26, 1993, p. 2D.

39. Charlotte (NC) Observer, October 30, 1993, p. 2A.

40. Product Safety and Liability Reporter (PSLR), July 9, 1993, p.737.

41. PSLR, June 18, 1993 p.650.

42. PSLR, June 25, 1993, p.676; also PSLR, July 23, 1993, p.785.

43. PSLR, Oct. 2, 1992, p. 1081.

44. 93 Law Week USA 282

45. From a summary of Senate Bill 667 from the office of Senator Rockefeller, dated March 26, 1993.

46. From a summary of House of Representatives Bill 1910 from the office of Representative Rowland, undated.

47. This is similar to the New Jersey rule, denying recovery where the claimant was found 51% liable for the harm. Cipollone v. Ligett Group, Inc., 112 S.Ct. 2608 (1992).

48. In his summary of the Bill, Rep. Rowland points out that "consumers will not have to pay exhorbitant prices for products to cover drugs or alcohol induced damages."

    Under the "socialization of risk" theory, the manufacturer/seller is held liable for the harm from his product, purchases liability insurance to cover himself and adds the cost of the insurance to the cost of product. By "passing along" this cost, all consumers of the product share in the cost of compensating for harm done by it. In his proposed national health plan, by increasing the tax on cigarettes, President Clinton would shift much of the burden of identified health costs from society in general to those who use the harmful product. One can see that H.R. 1910 would move in the opposite direction, in that all of society would share this cost of caring for those indigents who suffer harm from drugs or alcohol.

49. 93 Law Week USA 123

50. Wall Street Journal, September 24, 1993, p.B6.

51. 21 PSLR 968, September 24, 1993.

52. Id.

53. Id.

6 THE RELATIONSHIP OF TECHNOLOGY AND LAW IN CIVIL SUITS IN JAPAN

TSUNEO YAMADA
Science University of Tokyo, Noda, Japan

1 Preface

In many instances, the relationship between technology and law has been studied or considered in terms of "risk-conscious society and issues related to private law". Very recently, a symposium on the same theme was held by the Japanese-German Society of Jurisprudence. The main theme of the discussion at the symposium was liability for environmental damage under private law and product liability. Of the lectures given at the symposium, one which was most interesting to me was the lecture given by Professor Christian of Bremen University on the subject of "Safety Regulations for Products and Product Liability" subtitled "Process of Rationalization under Private Law", but I will not speak about it today. I will merely try to explain, by citing two or three examples, how the technology-related parts of civil suits are assessed in Japan, and through these examples, offer my personal opinion on how technology should be handled in the field of law.

2 Assessment related to technology in civil suits

2.1 Product liability

At present, there is no product liability law in Japan. Product-related liability has been handled as an issue of contractual liability or liability for an illegal act. The question of whether there has been any cause and effect relationship between negligence and the result would determine whether the manufacturer was liable or not. In other words, to hold the manufacturer liable, negligence on the part of the manufacturer would have to be proven one way or another. The bill for the introduction of product liability law in Japan, expected soon to be brought to the Diet, is believed likely to approve of defence based on development risk. I understand that in the unified law of the EC, the clause related to defence for development risk is an optional clause, and that this is going to be reviewed in 1995. Development risk means product risk which could not be anticipated at the levels of scientific or technological knowledge available when the product was placed in the distribution process. The question of whether to grant the manufacturer immunity for the reason of his inability to anticipate the risk with the

Structural Failure: Technical, Legal and Insurance Aspects. Edited by H.P. Rossmanith. Published 1996 by E & FN Spon, 2–6 Boundary Row, London SE1 8HN, UK. ISBN: 0 419 20710 4.

scientific or technological knowledge available to him at the time he placed the product in the distribution process is a political question, rather than a theoretical question.

Product liability is a rationale to hold the manufacturer liable for damages resulting from a product. It would be necessary to take this concept further and make it the manufacturer's obligation to try to make his product as safe as possible.

It appears that in Europe the matter of product liability is being discussed in terms of court's jurisdiction, service, examination of evidence and enforcement. It would be necessary to study further the technological aspect of the matter.

2.2   Intellectual property right
It is not only in the case of a court's decision relative to products that the technological aspect of lawsuits is not fully taken into consideration. In fact, it is usually in trials involving patents or copyright that the technological substance of the matter becomes an issue. In these trials, it is not infrequent that the parties do not make claims on points related to the subtleties of the technology because they do not want to disclose the content of the technology in court. The judge does not exercise his right to ask for explanations on those points, and passes a verdict without including such points in his judgment. As a consequence, the court's decision is extremely difficult for the technologists to understand. Even if they can understand it, they find it quite frustrating.

Let me show you fairly recent examples of courts' decisions.

In a case where the issue was whether the design drawings of a machine tool were objects of copyright, the court decided that the plaintiff's drawings were "drawings having a scientific nature" under Article 10-6(1) of the Copyright Law, on the grounds that the drawings had the aspect of expressing a technological idea acquired in the course of research and development, and that expression was believed to have originality. The court overruled the defendant's defence that "reproduction" did not apply here because machine drawings were not objects of copyright.

In the above case, the technologically important point is that the technological idea incorporated in the machine tool in question (a round bar correcting machine) shows up in the roughness of the surface finish. Even if the design drawings of the plaintiff and the defendant were both drawings of round bar correcting machines, it would be a legitimate claim to say that they were different objects of copyright if there were substantial differences in the diameters and surface roughness of the round bars, the objects of the designs. In the case cited, these points were not reflected at all in either the parties' claims or the court's decision.

In filing a second patent application on an invention which is closely related to the invention patented in the first patent application, the inventor or applicant may not refer to the patented technology of the first application when he writes claims in his second patent application, because he may think it is obvious. However, it would be meaningless if the applicant himself kept the relationship between the two inventions in mind, because the examiner does not take the first patent into account when he examines the second patent application unless the first invention is included in the claims of the second application. It is a question if these points should not be considered in courts' decisions regarding patents so that the decisions would be easy for the technologists to understand.

2.3 Employer's obligation to consider safety

In Japan, there are no legal provisions concerning the employer's obligation to consider safety, as there are under Article 618 of the Civil Law of Germany and Article 1157 of the ABGB. Therefore, in 1975, the Supreme Court of Japan, under a case law, recognized the employer's obligation to consider safety. Since then the number of cases in which damages are claimed based on the employer's breach of his obligation to consider safety has increased. Among these cases related to technology was a case involving the use of the chain-saw. Without going into the details, the purport of the judgement was that during the period of 1955 to 1961 it was natural that the employer could not adequately anticipate the nature and extent of vibration-caused impairments resulting from the use of a chain-saw or the like, and that therefore, the employer was not liable for damages for the ailment in question. The court further stated, "With respect to machines and tools whose necessity and usefulness are recognized because of social and economic development but which may involve potential danger, it is necessary to heed the possibility of any danger or damage that may result from their use, and take reasonable steps to prevent the occurrence of such danger when those machines or tools are used, rather than prohibit their use. However, if damage should occur when measures assessed to be reasonable in light of accepted social standards had been taken, the employer could not be regarded as having failed to fulfil his obligation to avoid consequences". Against this decision, one judge wrote a minority opinion, which is worth noting. He said "No matter how useful a machine may by, it is not right to belittle the effects of its use on the worker using the machine. The chain-saw was introduced without first making studies or investigations regarding its safety or safety standards for its use. Therefore, when it became possible to anticipate the occurrence of an impairment to the worker using a chain-saw, the employer is believed to have had the safety consideration obligation, even before he was able to establish corrective measures, to take temporary steps to prevent or abate such impairment by limiting the use of the tool. The Forestry Agency was in a position by around 1960 at the latest to anticipate the danger and therefore should have developed and implemented measures to limit the use of the tool as a temporary measure by around 1963 at the latest. However, the agency did not take any significant action until around 1965, thus failing to fulfil its obligation to consider safety. I believe that the state failed to fulfil its safety consideration obligation, and is liable for damages to the plaintiff".

When science and technology progress rapidly, the probability of problems like the one involving the use of the chain-saw occurring increases. It is believed that as technology becomes more sophisticated, the difference between "measures assessed to be reasonable in light of accepted social standards", and "measures judged to be highly necessary from an expert's viewpoint" will expand.

The obligation to consider safety concerns the safety of the lives and health of workers. Therefore, if the employer took the position that it should suffice just to take reasonable measures based on accepted social standards, the workers could not feel safe in performing their work. It may be a different story if workers are given the right to refuse to provide labor for reasons of the employer's failure to fulfil his safety consideration obligation. But under the current conditions in Japan where workers are not given such right to refuse to provide labor, I believe it is important to make it the employer's safety obligation to take "measures highly necessary from an expert's viewpoint".

There will probably be a growing number of cases in the future where workers are expected to work in areas unknown to science, such as outer space experiments. It would then be no longer possible to dismiss the issue of the employer's obligation to consider safety merely on the basis of the theory of assumption of risk.

3   Discussion

What I have discussed so far is only a small part of the relationship between technology and law that has come up in lawsuits in Japan.

Science and technology have progressed remarkably since the Industrial Revolution. Until the computer began to develop, society as a whole kept changing with the progress and development of science and technology, and therefore the disparity between the two (that is, the scientific and technological development and social change) is believed not to have been very substantial. However, since the computer began to develop, science and technology started to develop so rapidly that social change could not keep up with it, and the disparity between the two appears to have expanded. This disparity has caused various social problems.

These problems include not only disputes over product liability or intellectual property rights, which have been increasing significantly, and various issues concerning employers' failure to fulfil their obligation for safety considerations. There are also numerous other problems, such as the issue of environmental pollution, the issue of waste disposal, traffic accidents including accidents involving aircraft, satellites and space shuttles, damage from medicines and medical malpractice.

Product liability has been discussed in considerable depth, mainly among lawyers, from the viewpoint of international civil suits. The question of how technology should be handled legally is also beginning to be discussed. As far as the whole subject of scientific and technological development is concerned, however, only gene-related technologies have been brought under the scalpel from the viewpoint of ethics. The discussion has not yet begun to address the issue of how new accomplishments of scientific and technological research should be introduced to the world, giving full consideration to the future of human society. Engineers and scientists engaged in the development of new technologies keep developing new technologies to fulfil their insatiable spirit of inquiry and desire to create things.

The products of their work have made and will likely continue to make the lives of the people on earth more convenient. However, this is not necessarily all good. The use of new technologies has created and will probably continue to create many harmful results. If this is so, we must check, and correct or take corrective action for the harmful results of the use of a new technology before that new technology is introduced to the world. It appears that consideration in this respect is beginning to be given to measures for the protection of the environment, but it is not sufficient at all. It would in no way be too late if we examined from various angles the effects that a new technology might bring on our society and took corrective measures wherever such measures were believed necessary, and then introduced the technology to the world. Everyone engaged in research and development should recognize the importance of this.

That means that it is necessary for international treaties to be concluded and for countries to reorder their laws. Not only technical and legal experts but experts in all sectors should sit at the same table and study this matter deeply and from a broad perspective. The court passes its judgement on what action should be taken on a problem after that problem occurred. This is natural because "there is no trial where there is no complaint", but legal scholars can think about preventive laws. It is extremely important that the legal community contributes in the efforts to examine the adverse effects that scientific and technological advancements may bring on society, make adjustments and take corrective action before the effects actually occur. It would also be essential, considering the possibility of damage occurring because of those adverse effects, to reorder the insurance system to provide for relief of the victims.

In Japan, work to make a product liability law has finally started, and the law is expected to be brought into effect in 1994. A product liability insurance law is still being studied by insurance companies, and no specific bill has been presented for study yet. It appears that the concern that the enormous amounts of compensation for the kind of punitive damages granted in America might lead to rises in product prices because of high insurance premiums is putting the brakes on the preparation of a bill. It is also necessary to study all possible measures to minimize the chance of damage resulting from products. One such study may be to probe into the possibility of an ombudsman system for science and technology.

7 THE STATUS OF PRODUCT LIABILITY IN POLAND

KRZYSZTOF CZERKAS
Failure Analysis Associates, Gdańsk, Poland
and
ELŻBIETA URBAŃSKA-GALEWKA
Technical University of Gdańsk, Gdańsk, Poland

1 Introduction

Product liability, a concept familiar to professionals and even ordinary people in democratic, market-oriented, countries, gradually become a new, to some extent strange and exotic, but - fortunately - unavoidable for a limited period of time, feature of the present Polish economy. As a heritage of the past economic system of central planning management, Poles have adopted bad technical and insurance laws, bad technical standards, and, what seems to be the most important, a bad idea of how things went wrong.

As the means of production have been the property of the State, the assessment of lifetime and the problem of durability in every field of technical activity in Poland had been a theoretical subject rather than a practical one for many years. With the lack of any relationships between work done and money gained, and with the domestic market almost empty of first-class goods, Polish economic leaders were not interested in the development of any economic, technical or legal moves in the field of quality control and product liability. Additionally, the irresponsibility at practically every stage of decision making (planning), its implementation, (designing) and execution (production) in central management, destructively influenced the quality of final products thought fit for the domestic market.

Since 1989, the process of replacing the old economic system with a new, market-oriented one has been in progress and trade has become of paramount importance, but few changes have been noticed in the field of product liability. To make matters worse, Poland has lately been flooded by imported products of which the quality has been below internationally accepted standards. Lack of specific laws and regulations, lack of trade experience amongst importers, as well as, sometimes, lack of common sense, led to a situation where firm and quick steps should be taken immediately to remedy the situation.

Similar situations have arisen on the Polish investment and real estate markets. In recent years, we have witnessed some typical instances of negative results associated with a lack of legally backed responsibility at particular stages of the investment or maintenance process. Although there are a number of contractors and sub-contractors involved in the process, the investor, often identified with the user of a structure, is the only one who suffers a loss from any error, negligence, or carelessness committed.

Structural Failure: Technical, Legal and Insurance Aspects. Edited by H.P. Rossmanith. Published 1996 by E & FN Spon, 2-6 Boundary Row, London SE1 8HN, UK. ISBN: 0 419 20710 4.

2 The legal state of product liability [1]

In Poland, action concerning product liability caused by a defective product is based on The Civil Code (Articles 556-582), dated April 23rd, 1964 [2]. In general, The Civil Code is based on the "negligence in tort" rule. There have been several modifications of the law between 1964 and 1990, but only when matters become very urgent (among others, the law concerning free land turnover). It was realized that further corrections, particularly the adaptation of the law to the changing economic conditions and regarding Poland's future access to the European Common Market, to standards obligatory in Western Europe are necessary. Such a proposal has already been prepared by the Civil Law Codification Board of The Ministry of Justice. It is in a package of 55 bills presented to Parliament by the resigning government of Ms Hanna Suchocka. Undoubtedly, its passage into law will be delayed as the new government, under the leadership of Mr Waldemar Pawlak, has temporarily withdrawn the whole package from Parliament.

2.1 The manufacturer's liability
In Poland today, the buyer of a defective product, such as a floor finish which causes a long-lasting allergy or a lamp which could become the cause of a fire, will receive damages from the manufacturer only when the consumer proves the manufacturer's guilt infringement of the production technology, standards, etc. The legal base for claims in such cases is fundamental for delict liability Article 415 of The Civil Law stating laconically, "Who did to somebody a damage through his fault, he is obligated to compensate for it."

In practice, it is extremely difficult to prove the manufacturer guilty of a defective product at present. A consumer, however, is protected by administrative law regulations dating from 1961 in regards to standardization and product quality law which impose on the manufacturer the obligation of following special requirements and standards, which are now obsolete and do not match modern standards.

It is assumed all over the world that the useability of a product is a matter of agreement. The legislator only determines those product features which are directly related to its safe use. These are the so called critical features. Whether or not a product is in good taste, is durable, or more or less efficient is not in question, but whether it will be a safety or health hazard to its users or other people, the environment, animals, and salubrity of crops. These critical values are determined by laws and executive acts. If the product does not meet these laws and regulations, it will not be permitted into the market, and, therefore, not be sold.

2.2 The seller's liability
There are two concepts in The Polish Civil Code Law: "fault warranty" and "guarantee".

Warranty is the seller's liability for physical and legal shortcomings of a product sold, independent from guilt, existing ex lege with the possibility of modification by the agreement of both sides. The right of warranty does not exclude a buyer's compensation claims by the right of non-performance or improper performance contract nor delict claims if the salesman's behaviour is illegal, e.g. selling of dangerous products with faults. Each seller is subjected to warranty, independently of whether he was the manufacturer of the product and he contributed to the fault or he was aware of the fault or could have detected the fault after a closer inspection. A seller is relieved from liability by the title of warranty if the buyer was aware of the fault at the moment in which he entered into the contract.

Guarantee represents the seller's (or manufacturer's) commitment to physical fault removal of products sold or to product replacement if the fault is discovered within the period in which it is guaranteed. The granting of a guarantee demands a written form. If not stated otherwise, a seller is liable by the title of guarantee only when the fault occurs for reasons inherent in the product sold.

In view of difficulties concerning the interpretation of legal principles included in Articles 556-582 of The Civil Code, The Supreme Court passed a resolution of guidelines addressing the matter of law interpretation and juridical proceedings in warranty and guarantee cases [3] on December 30th, 1988. The resolution includes nineteen appropriated guidelines and their comments. It was also stated in the resolution, among others, that:

- while assessing physical faults of an object, the criterion of functionality, including the useability of a product and its assignment according to the sales contract, should be given priority in regards to the standard-technical criterion,
- the obligated party, by the title of both guarantee a warranty is free to chose the form of liability fulfilment towards the buyer either by repairing the faulty product or replacing it with a new and fault-free one. Repairing the fault in order to restore the product to its proper use in accordance with its purpose should be possible and should not turn out to be troublesome for the buyer.

Those two parts of The Supreme Court's decision seem to point out that the legislator did not pay much attention to the proper protection of the buyer's prime interest in buying fault-free products.

It is worth stressing that, both Civil Code regulations and The Supreme Court decisions were issued in a period when the fundamental citizen's law, including those of the consumer, were subjected to changes.

Despite fundamental changes in Poland's system, which took place in the year 1989 and consisted of transforming the centrally planned and imperative-distributive national state economy to a market-oriented economic system based increasingly on private capital, the above quoted principles and law directions concerning the protection of the consumer interest have not been verified yet by Parliament. If the previous law code had been adapted to fit the previous social-economic conditions in Poland, it would be obsolete, even "dead" in the present Polish social-economic situation as it is still mainly concerned with the state-controlled economy. According to official data from the Polish government, the share of the private sector in production, trade, and service is about 45 per cent, while trade, in the private sector, is about 90 per cent. Just these two factors clearly demonstrate how the still obligatory law regulations are inadequate under the present conditions.

The existing state legislation indirectly favours the appearance of an increasing number of dishonest sellers, who, involuntarily or, what is worse, voluntarily, do not perform their obligatory duties concerning liability for products and services sold. It causes an ever increasing number of conflicts between buyers and sellers and, in most cases, the buyers are at a disadvantage.

3 Proposals for changes in the law [4]

3.1 Liability of a manufacturer

Proposals for changes in civil liability for his or her product are in complete agreement with the principles of liability given in Directions No. 374 from the year 1985 of The Council of The European Community. The fact that Polish membership in the European Common Market continues to remain a future possibility and the quality level of national production is poor. It is proposed that normalisation and harmonisation, in accordance with the mentioned directives should be introduced into The Polish Civil Code. After Poland is granted membership in The European Common Market, a law about compensation liability for a product, modelled after those European Common Market countries, should be issued.

Liability for damage caused by unsafe products is to be settled according to Articles 434.1 and 434.2. They will be included into existing regulations for liability for animals and objects and for liability for damages committed by functioning of enterprise and means of communication. The intention of the originators of those regulations has been that the new regulations, together with above settlement, should create a harmonious group of regulations concerning the risk of the professional manufacturer. The essence of the proposed regulations is the producer's liability for damages resulting from shortcomings of the basic feature of the product safety based on the risk principle, which is different from liability based on the guilt principle.

In accordance with the proposed Art. 434.1 of The Civil Code, a manufacturer who introduces a product on the market within the range of his professional activity, will be liable for damages inflicted upon a person or his property as a result of normal and proper use of the product unless the damage occurred through some fault of the user or another's fault due to circumstances outside a person's control. Un-processed agricultural products and those of "raised animals" are distinctly excluded from this regulation.

The Authors of the amendment wish to point out that, as a result of the standardisation, manufacturers will be charged with the risk of introduction of new technical solutions (technological progress). It should also entice manufacturers to properly inform a user about the conditions of its use and maintenance of the product, thus giving the user appropriate service instructions.

It is proposed that the same liability should apply to the materials and the product component manufacturer to the one who placed his label on the product, the product salesman, and the importer who introduced foreign made goods into the markets should be made jointly liable.

Given such a regulation, it is extremely important that in case of damage, a manufacturer (importer, seller, etc.) will have the chance to provide any evidence or facts that which exclude him of guilt. If he fails to do this, then he will have to pay the damages. The scenario will be quite different from the present situation where the customer is expected to prove the guilt of the producer and, when he is unable to do that, is not compensated.

However, important restrictions with respect to the range of liability will have to be considered. Product liability will not include the product itself and advantages which the customer could gain from the product by using it. Damages might be claimed only when the value defined by The Ministry of Justice is exceeded. Therefore, damage cases involving small-scale compensation have been excluded.

The proposed regulations concerning liability for damage caused by a faulty product assume that a manufacturer is liable ex lege and is obligated to provide a safe product. These regulations are being created mainly for the benefit of a customer as the final user of the product.

3.2 Liability of a seller

Article 560 of The Civil Code, which is mandatory at present, implies that the buyer is entitled to withdraw from the agreement or demand a price reduction. The buyer, however, can not withdraw from the agreement if the seller declares his willingness to immediately replace the faulty product or to immediately repair the defect. This means that a buyer may have to accept a repaired product.

The authors of the prepared amendment to The Civil Code suggest that the repair alternative be excluded. This solution, which is disadvantageous for the consumer, was widely criticized. Therefore, a seller might "rescue" the agreement only when he has a faultless replacement in stock at his disposal. If the faulty product has already been replaced by the seller and the product is also faulty, the buyer can refuse to accept replacement and withdraw from the agreement and demand that his money be refunded. At present, because of such a restriction, the replacement "game" might last forever.

The part of paragraph 3 of Art. 560 of The Civil Code which deals with the price reduction as a result of the product sold being at fault (when a regulation is enforced which defines the price for the product of a given sort or class, a buyer is to pay for the faulty products according to this regulation) is obsolete and will be deleted. A new paragraph 4 is to be added to art. 560 of The Civil Code which strictly states that "if the seller made a replacement, he should cover the costs incurred in connection with this replacement and be borne by the buyer" (e.g. transport cost).

The content of various articles in The Civil Code is to be amended towards decreasing strictness with respect to a buyer and by making clear and simplifying stipulations, the lack of which, was the reason for numerous disputes and doubts.

The authors of the amendments claim that from a legal point of view, changes in legal solutions concerning warranty for errors have far reaching consequences. Thus, in all regulations concerning warranty, the definition "product" was replaced by the more general term "object for sale". This accounts for not only for products, but also property including bonds, stocks, and shares. As a result of supplementing the definition of legal error, it will include, apart from the situation where an object for sale is the property of a third person or is charged by the rights of the third person, also those cases in which an object for sale is restricted by the rights of the third person. Moreover, it will be provided with the statement: "in case of selling legal titles, the seller shall be liable for the existence of any rights or claims".

Other amendments of substantial character will enhance the position of the buyer of a legally defective article. A buyer will be eligible to claim his rights by the title of warranty in case of a legally defective product, even if a third person will not oppose him in court with the warranty for the object of sale. The buyer of such an object must not be compelled to be constantly threatened that the person may confront him in the future.

It is also proposed that, in view of safety enhancement the expiration period of the liability by the title of warranty against defective property should be extended up to three years, starting from the moment when the buyer learns about the fault. At present, this expiration period is one year regardless of the kind of article for sale.

The regulations concerning guarantee require numerous changes in the present regulations to be made. The legislator speaks, as a rule, about a "seller" being obliged due to the guarantee and a "buyer" as having the right of guarantee. The use of those expressions is incorrect since the one who is obliged to guarantee is usually the manufacturer and not the seller, and the one who is entitled to guarantee may not be the buyer but the one who owns the object of guarantee and the guarantee documents. As a result the definition "seller" should be replaced by a "guarantor" and a "buyer" being entitled to the guarantee.

The introduction of obligator guarantee will constitute a very important change in law practice with respect to customer's interest, this legally obligatory guarantee shall be for all professionals regardless of their private or official status (at present, executive regulations obligatory within this range concern only state-controlled units). A seller will be obliged to provide a buyer with a guarantee document. If a seller refuses such a document, guarantee will focus legal liability exclusively on him.

4 Civil Engineering Code [5]

Polish Building Law has been obligatory in Poland ever since its proclamation on October 24th, 1974 [6] until the present. It has been severely criticized since the middle of the eighties, and all the changes and supplements introduced (ten times) in the meantime, by mode amendments made the situation worse by decreasing the compactness of law solutions. As a result, a body of Law Legal Regulations has been drafted recently. The Polish Parliament passed a resolution concerning the Building Law on March 20th, 1993. The Polish Senate accepted the resolution including, on April 23rd, 1993, numerous amendments. Unfortunately, the Polish Parliament was dissolved unexpectedly just one day before the second voting on the Regulations.

The purpose of the Building Law was to adapt Polish architecture and civil engineering to market/oriented economic conditions. However, it was worked out at the time when new market oriented economic patterns had not yet been developed such as functions, professions, and working and operating in the building sector.

First of all, the drafted law provided for the enhancement of the protection of the social interest in the building process by increasing the liability of the participants in the process for the proper performance of their tasks. The draft determined precisely the liability of the parties which independently perform their tasks in the building process. It specifically addressed the liability of the designer and the fulfilment of his authorial survey, the construction site foreman and the investment supervisor. According to the law drafted, the investor was the only one liable for the organization of the building process which not only includes the design and its performance, but also the acceptance of the construction work. The investor was free to, but did not have to, choose either the investment supervisor or the designer to fulfil the authorial survey. However, with the intention of protecting the public interest, the possibility of making an investor liable for a supervisor appointment is under consideration. On the other hand, however, it appears that the investment supervision defined as above too one-sidedly exposes the role the supervisor plays as an investor's representative on the building site. In the new economic system, it will increasingly become the case that the investor will not be the owner of the building. Such a case might occur when the demand for cheap housing arises and it pays to build flats as cheaply as possible without paying special attention to their quality. In this case, the investor's supervision would not ensure the proper protection of interests of future owners or flat renters.

The draft of The Building Law Regulations also defined the liability of the designers that they would be considered fully liable for the project. They would have the right to visit the construction site and make appropriate notes in the daily report of the construction project. Unfortunately, no system of protection for designers and building offices exists in Poland. Prospective legal and organizational activities should address this topic in the near future.

In order to increase the level of liability of each subject in the building process, changes in The Civil Code are necessary. The Civil Code regulates matters concerning construction work related warranty and matters concerning construction works. In case of damage, the present version of The Code requires contractor's guilt to be proved and the warranty period is for three years. It places the investor (user) into a position worse than that comparable in the western countries where the principle of the liability of the contractor for the defects is legally regulated and the warranty period is a minimum of five years, which sometimes extends even to ten years. These conditions, in terms of the contractor's liability and the fact that for defects to occur this frequently, the field of construction associated with high repair cost force the contractor to search for protection for himself. Subsequently, insurance agencies imposed their own requirements concerning the quality of work and materials, and when inspecting the fulfilment of these requirements they accordingly adjusted the insurance premium in order to protect their own interest. Such a sequence of motivations and actions induced by the various facts leads to spontaneous inspection mechanisms and have the effect of increasing quality without a state administration representative's engagement and interference.

The present rules of The Civil Code concerning liability and warranty of construction works do not offer appropriate requirements and motivations for achieving high quality in construction works and also do not sufficiently protect against the consequences of defects and shortcomings with respect to building performance and against using materials of poor quality. This situation demands proper changes in The Civil Code.

In countries with market-oriented economies, standard regulations addressing the relation between the participants of the investment and construction process have been developed. Illustrative examples include regulations about the selection of construction work contractors (tenders), about the kinds and general conditions of contracts including ways of giving notices of termination, about the principles of defining the mode of payment for construction works, settlement of charges, price actualization about principles advancing conventional penalties, conflict resolution, acceptance conditions, etc. The Civil Code comprises only general regulations concerning construction work contracts. According to The Code, the parties are free to define their reciprocal commitments. In such a case, the way of pre-contractual negotiations in which the parties involved reach consensus, becomes of principal importance. It also forces the parties to precisely record in the contract all the reciprocal commitments without the possibility of recalling other settlements in other regulations or documents. This situation would be disadvantageous, and changes should be introduced following along the lines of countries with market-oriented economies.

5 The Polish insurance market

5.1 The present system of insurance in Poland
During the 1980s, Poland faced a period which has been generally considered as the "Lost Decade", in which economic indicators stagnated (or went into reverse) and social

problems grew larger and larger. In the 1990's, Poland, already accounted for slow, but still stable, economic growth.

In 1994, the insurance market in Poland is represented by active operation of about 20 insurance companies. In fact, more than 50 per cent of Polish insurance activities relate to three of them, namely: Powszechny Zaklad Ubezpieczen S.A. (P.Z.U. S.A.), Towarzystwo Ubezpieczen i Reasekuracji "WARTA" S.A., (WARTA S.A.), and Sopockie Towarzysto Ubezpieczeniowe "HESTIA" S.A. (HESTIA S.A.).

Risk associated with product liability has not been at the forefront until 1994. Up until now, claims considering product liability in Poland have been dealt with in accordance with the notion of "negligence", one of the basic tenets in The Civil Law, since, in Poland, there is no law dealing specifically with product liability. Underwriting results for Product Liability insurance within Poland itself have been quite stable and varied within a fairly narrow range. Recently, however, the demand for strict liability formulated along the lines of that adopted in The EC has been growing. At present, this issue is being debated at the meeting of The National Insurance Council, which has just been founded in Warsaw.

Up to now, the approach employed in the field of insurance supervision in Poland represents a mixture of the approaches taken in other countries in Europe. The insurance conditions and, in some cases, the premium of products to be insured must be approved by the supervisory authority.

5.2 Consumers and insurers
The foreign trade and contracting partners have demanded the Polish companies to take out product liability insurance. Only the insurance company Towarzystwo Ubezpieczen i Reasekuracji WARTA S.A. (TUiR WARTA S.A.) has been allowed to offer General Conditions of Insurance since World War II. Baltyckie Towarzystwo Ubezpieczen i Reasekuracji HEROS S.A. (BTUiR HEROS S.A.) have introduced similar offers for a year and Sopockie Towarzystwo Ubezpieczeniowe "Hestia Insurance" S.A. (STU "HESTIA Insurance" S.A.) is just now preparing General Conditions of Insurance for product liability. Exclusions, e.g. concerning bodily injuries and property damages caused by conceptional and constructional defects and omissions, apply if, as a result of this, an entire production process has become defective. The lack of experience of Polish insurance companies and the shortage of Polish legal regulations have caused exclusions of the kind mentioned above.

Fortunately, changes in the Polish economic market have given rise to a new kind of insurance. Four Polish insurance companies, Towarzystwo Ubezpieczeniowo Reasekuracyjne POLISA S.A. (TUR POLISA S.A.), Zaklad Ubezpieczen HESTJA S.A. (ZU HESTJA S.A.), Towarzystwa Ubezpieczeniowe COMPENSA S.A. (TU COMPENSA S.A.) and Pomorskie Towarzystwo Ubezpieczeniowe (PTU GRYF S.A.) offer liability insurance to civil engineers. This concerns responsibility for technical and human errors of persons insured or his workers as well as for direct consequences.

Insurance companies have utilized data concerning the probability of occurrence of construction damages to assess the level of risk. The data mentioned were taken from the Institute of Construction Technology (ITB) and The Polish Union of Civil Engineers and Technicians (PZITB). These data are derived from detailed records concerning construction damages and diasters which occurred in Poland between 1961 and 1985. Rational analysis of these damages and disasters has yielded the following distribution of the types of processes in Poland:

- designers	4 - 20%
- building contractors	50 - 80%
- producers of construction materials and prefabricated elements	4 - 16%
- users	4 - 20%
- others	2 - 6%

The types of insurance described are a novelty in our country and are of really significant importance. They not only help to prevent construction companies from suffering financial crash as a result of damage compensation, but more important, they exert a great influence on society to change her attitude towards work and help to improve the quality of the work.

References

[1] Lewandowska I., "Bubel niebezpieczny", Rzeczpospolita, No. 251 (3595), October 26, 1993.

[2] Ustawa z dnia 23.04.1964 r. "Kodeks Cywilny", Dziennik Ustaw Nr 16, poz. 93.

[3] Uchwala pelnego skladu Izby Cywilnej i Administracyjnej Sądu Najwyższego z dnia 30.12.1988 w sprawie wytycznych w zakresie wykladni prawa i praktyki sądowej w sprawach rękojmi i gwarancji, Monitor Polski Nr 1/1989 r., poz. 6.

[4] Lewandowska I., "W interesie konsumenta", Rzeczpospolita, No. 253 (3597), October 28, 1993.

[5] Biliński T., Proposal for legal regulation changes in civil enigneering code, XXXIX Konferencja Naukowa Komitetu Inżynierii Lądowej i Wodnej Polskiej Akademii Nauk.

[6] Ustawa z dnia 24.10.1974 r. "Prawo Budowlane", Dziennik Ustaw Nr 38, poz. 229.

8 PRESENT-DAY SWISS LEGISLATION ABOUT PRODUCT LIABILITY

TITUS PACHMANN
Frick & Frick, Attorneys at Law, Zurich, Switzerland

1. In Switzerland, product liability has not created a great
    stir in jurisdiction to this day. Since 1923, only a few
    federal decisions about typical cases of product liability
    have been issued. Nine <u>leading cases</u> are a modest output
    for seven decades.

The first of those nine cases was about a shoe-polish,
which induced an eczema in the year 1923. Further, there
was a climbing-belt (1938), which was torn apart and made
an overhead cable electrician fall down. Another case was
about a growth expedient for plants (1961), which was
applied to vines and made them wither. A deep-fat fryer
became the cause of a fire (1964). A clothes detergent
developped gases and also caused a fire (1970). A mechanic
repaired a car not carefully enough and the accelerator got
stuck, which lead to an accident (1980). The detergent of a
cleaning machine was diluted with petrol, and the gases
caused fire (1984). The shaft frame case also of 1984 was
about a concrete frame weighing 600 kg, into which carry-
ing-straps were cast. When the concrete frame was hoisted,
one of the straps of steel broke and the frame severly
injured a worker. The last renowned case was about folding
chairs. A dentist had furnished his waiting room with
Italian chairs, the rivets of which were too weak. A
patient, under which such a chair broke down, suffered a
spine damage (1985). After all, problem cases have
increased distinctly in the last twenty years.

I would have liked to present you at least one new attrac-
tive Swiss case about product liability. Unfortunately, I
have to disappoint you - the last such newspaper report of

Structural Failure: Technical, Legal and Insurance Aspects. Edited by H.P. Rossmanith. Published
1996 by E & FN Spon, 2-6 Boundary Row, London SE1 8HN, UK. ISBN: 0 419 20710 4.

August 1993 with the headline "splinters of glass in beer" did not concern Swiss beer, nor deliveries to Switzerland. Hence, you still may order and drink your beer in Zurich or Geneva without worrying.

Now, why are such federal decisions in cases of product liability so rare in Switzerland? Two reasons can be named: on the one hand the companies' very high insurance coverage and on the other hand the very high percentage of compromises in cases of damage.

That results in Swiss' jurisdiction showing liability cases about ski-runs or incorrect credit information of banks rather than typical cases of product liability.

2.   On January 1st, 1994, that is in 43 days, the federal law about product liability of June 18th, 1993, will come into force in Switzerland. Linked to it, an amendment of the federal law about the safety of technical installations and instruments will also come into force. These two laws will set new rules in Swiss product liability. But why this legislative initiative?

You all know that within the framework of the European Economic Union and within the framework of the European Economic Area, the product liability guideline of the EEU is decisive. According to this guideline, product liability regulations corresponding to it have to be included into national law. Within the framework of the EEA negotiations, Switzerland was required to assume the present law of the EEU to a large extent. It was therefore intended to include also the product liability guideline into national law. On December 6th, 1992, the Swiss people refused to join the European Economic Area. So the Swiss legislative body

decided to enact the prepared federal law about product
liability in a slightly modified form on its own initiative, in order to be compatible with the European Economic
Union and the European Economic Area in this respect. In
consequence, the mentioned federal law about product liability was enacted and will come into force.

3. Product liability signifies the liability of the producer
   for damages resulting from faults of his product. This is
   an extracontractual liability. Hence, a liability that is
   applied independently of there being a contract between
   producer and aggrieved party or not.

   The Swiss legislation tends to assume the EEU guideline
   about liability for faulty products of June 25th, 1985,
   85/374 EEU. It only differs from it in one point, to which
   I will come back later.

   "Producer" is the producer of the final product, the producer of a primary component or the producer of a partial
   product. "Producers" are further persons who pass themselves off as producers by attaching their names, their
   trade-marks or another sign of identification to the product. Furthermore, the importer is also classified as producer. Finally, the retailer is considered as producer when
   the producer himself or the importer can not be established. According to these regulations, several persons can
   be liable jointly and severally. The producer of a partial
   product can release himself if he proves that the fault lay
   in the construction of the product into which the partial
   product was incorporated.

Covered <u>damages</u> are death or physical injury as well as
material damage suffered by the <u>consumer</u>. Material damage
applies to articles of property in the <u>private, non-commer-
cial</u> sector. In the area of commerce, product liability is
not applicable. The aim of product liability is consumer
protection. The own share amounts to SFr. 900.-- (corresponding approx. Ecu 500.-- according the EEU guideline).

A <u>product</u> is every movable object, even if it forms part of
another object. Also electricity is considered a product.
Here, however, the only possible damages are too high
voltage or too high intensity of current.

The law is not applicable to <u>nuclear accidents</u>.

A central regulation of the new legislation concerns the
<u>concept of faultiness</u>. A product is faulty if it does not
provide the safety one can expect taking into account all
circumstances. As a yardstick, the <u>objective level of
expectation of the general public</u> is taken. The presentation of the product has to be taken into consideration,
such as imprints, instructions, advertising and other
statements about the product. Furthermore, the use of the
product that normally can be imagined has to be considered.
This use has to be interpreted more broadly than the designated use. For example, it has to be reckoned with paper
products' being used for making a fire. Finally, the point
of time of the product's being put into circulation has to
be taken into account. If the safety requirements have been
reinforced after that time, they are not decisive for that
product. Also a product does not turn faulty just because
there have been produced better ones after it.

4. Product liability does <u>not depend on the principle of personal guilt</u>. The aggrieved party has to prove the damage, the fault of the product and the adequate causality between fault and damage. To relief himself, the producer may bring forward the following evidence:

    — He did not put the product into circulation.

    — The product was not faulty when it was put into circulation.

    — The product was not intended for sale or for other marketing purposes with economic aims.

    — The product meets binding sovereign regulations.

    — The fault could not have been detected with the level of knowledge in science and technology at the time the product was put into circulation.

In contrast to the usual <u>limitation period</u> in Swiss law concerning extracontractual damage of one or two years, the new federal law contains a limitation period of three years and regulations for a forfeiture period of ten years. The period of forfeiture can only be interrupted by instituting proceedings.

Switzerland has abstained from extending product liability to agricultural natural products and to hunting products. Also, no limitation of the liability sum was established.

In general, the Swiss legislator has declared the current law of extracontractual damage to be supplementary applicable. That leads in one small point to a contradiction to the EEU guideline. According to Swiss law, the judge may

reduce the compensation to be paid in case the liable
person is in serious financial difficulties (art. 44
para. 2 Swiss Commercial Code). In the EEU guideline, no
such regulation is provided. Therefore, this is a <u>contra-
diction to the EEU guideline</u>. In my opinion, however, the
contradiction is of minor importance, since that article is
applied only if the damage was caused neither intentionally
nor by gross negligence and if in addition to that full
payment would bring the liable person into serious financial difficulties.

Here, it has to be added that in Switzerland a <u>revision</u> of
the law of extracontractual damage is in progress. One of
the propositions of the reform committe is to extend product liability from material damage of consumers to all
kinds of material and property damage. Therefore, it could
be conceivable in the medium-term that in Switzerland also
commercial enterprises could refer to regulations of product liability. That would be an extension also with regard
to the EEU regulation, but it would be an extension in
correspondance with EEU law.

In its <u>interim regulations</u> the new law provides that it is
only applicable to goods that were put into circulation
after the law came into force.

5. <u>All in all</u>, it can be ascertained that the law takes up the
    very strict requirements concerning construction and end
    control that are already in force today in the area of
    Swiss extracontractual damage (federal decision BGE 110 II
    456). It is a rule even today that the parts of a product
    that are visible and easy to control have to be examined by
    the seller if the product holds certain dangers, if it
    comes from a producer whom the seller does not know or if

the seller has reason to doubt the quality of those products.

6. Now, what is the significance of the new legislation for
    Swiss exporters and for importers of Swiss goods especially
    in the area of the EEU, that is when a product is fabri-
    cated in Switzerland and then exported to an EEU- or an
    EEA-state.

    According the EEU guideline, the importer who imports goods
    from Switzerland to an EEU- or EEA-state is liable besides
    the producer for the damage the product causes. However,
    the importer is free to make an arrangement with the pro-
    ducer by which he is released from any liability risks.

    The aggrieved party has no possibility to sue the importer
    for the suffered damage if the involved states are members
    of the European Agreement about judicial competence and
    enforcement of judicial decisions in civil and commercial
    cases of September 16th, 1988. The agreement is called
    Agreement of Lugano by its place of completion
    (SR 0.275.11). At the moment, the Lugano Agreement is valid
    for France, Great Britain, Italy, Luxembourg, the Nether-
    lands, Norway, Portugal, Sweden and Switzerland. In addi-
    tion, the EEU states are connected in the same way in the
    so-called Agreement of Brussels.

7. As a conclusion of the new law of product liability in
    Switzerland it is recommendable for producers to observe
    the following:

    - The products should be examined under the point of view
       of product liability.

- The producer should in any case save planning, construction and fabrication documents of his products as proofs.

- In advertisements, instructions and also on the packaging the dangers of the product should be indicated.

- Recourse claims of the possibly liable persons or companies among each other can be limited by contract. I think about the relationship of producers/subproducers/ suppliers/importers. They can make agreements among each other to exclude/broaden liability. Suppliers and subproducers can make agreements to exclude liability as a result of a fault.

- The insurance coverage of the enterprises should be checked.

From the point of view of the consumer, the matter is very simple. The new law about product liability in Switzerland is stringent law and therefore applicable without exception.

8. In the Sixties, Robert F. Kennedy pronounced the right of the consumer to safety, faultlessness, information and free choice. Today, also in Switzerland every enterprise should observe the urgent advice:

"Management should give safety the same priority that it gives costs, delivery and functional performance."

9 PRODUCT LIABILITY - IN CASE OF A LICENSE?

PETER REVY von BELVARD
European Patent Attorney, Wil, Switzerland

Introduction

Ladies and gentlemen, please do not expect me to deliver a lecture which will make you masters of this rather difficult area. There is a vast literature on this subject so that all I can do is give a general overview of the problems you may encounter.

It is, therefore, perhaps surprising that license contracts are often "home-made" by persons generally unfamiliar with the questions which may arise.

Types of licenses

When speaking about licenses, we should define what a license is or can be. A license is a permission granted to another person to make use of something which is protected either by protective rights, which means Patents, Utility Models, Designs, Trademarks or the like, or by secrecy as is the case with Know-How. A speciality is software licenses. Often some of these types are combined.

Types of risks

In all those license agreements, there are a number of risks which can result in tortious liability.

1. The most common crucial point is the question of the validity of the licensed protective rights, i.e. of Patents, Utility Models, Designs and Trademarks.

2. But also in the case of a Know-How license there exists the problem of whether the protection is actually given, and this means whether it does contain secret material of sufficient importance.

Structural Failure: Technical, Legal and Insurance Aspects. Edited by H.P. Rossmanith. Published 1996 by E & FN Spon, 2–6 Boundary Row, London SE1 8HN, UK. ISBN: 0 419 20710 4.

3. Another question, which is not commonly treated in license agreements, is the problem of being able to use the licensed technology freely without infringing the rights of third parties. This can be a stumbling-block in all types of license agreements.

4. Particularly in the case of licenses on Trademarks and Know-How, there is the question of quality of the final product. If it does not have the expected quality, it might diminish the reputation of a Trademark. In addition, tortious liability may arise, in which case the question of quality will split up into

   a)   the liability for a technical failure of the licensed and transferred technology or information, and this will be the case particularly in Trademark agreements where it is usual for the licensor to retain a right of quality control.

   b)   the liability for a technical failure of the product manufactured by the licensee. It is not obvious that the licensor cannot be liable in this latter case, although he may not have been directly involved with the production. For some court decisions show that a license is sometimes considered as a form of technical assistance to "selling" a dangerous product. Moreover, Belgian law holds that a licensor, unless the obligation is disclaimed, guarantees that a patented invention is feasible, and also other countries imply an obligation on the licensor to guarantee technical performance.

Situation in different countries

To render the situation more confusing, we have to take into account that we will find quite different legal situations in different countries.

The United States is the classic country for spectacular liability cases. But regarding the present legal status in the United States as well as in Canada I refer to the lectures of my colleagues Mr. Peters, Ross, Kirkwood from the United States and Mr. LeMay from Canada. Here, I wish only to point out that the liability of the licensor will be supposed especially in case of Trademark licenses. This may be intelligible, if one considers the licensor's possibility of quality control which is inevitably associated with this type of licenses. But liability is a growing concern like a growing plant gradually entangling other types of license contracts.

Thus, the literature on court cases supports the opinion that liability results from a misperformance of the contract and the licensor can be charged with misfeasance which resulted in injury to a third person. Whether a liability exists will depend to a large extent on the terms of the licensing agreement and has to be examined from case to case. But lack of safety of a patented product might also be the reason for a licensor's liability.

There are some countries, such as the **Andean Common Market** or **South Korea** which require that "Every license contract shall contain provisions which assure the quality of products and services provided by the owner of the license." And if such a provision is contained, a liability of the licensor cannot be excluded, unless appropriate sophisticated clauses are inserted.

While the situation in **Canada and Japan** is quite similar to that of the USA to the point where those countries can be considered as the "dangerous countries", this shows clearly what we have to expect from the corresponding **EC-rules**. According to these rules, the producer is liable even if his product is not considered as being defective from the technical and scientific points of view at the time of putting it onto the market. And a "producer" is not only the producer of the final product, but also of an intermediate product and every person whose name or trademark is on the label! It is significant that a similar definition is in the **Austrian** "Produkthaftungsgesetz". However, while the effects of these rules cannot yet be fully estimated, there still exists a lot of nonconformity as well as some confusion, in the different countries of the EC.

Although it is common in all countries that the claimant has the burden of evidence for defects and causality of the damages and the defects as well as the fact that the defect was present at the time when the defendant issued the product, there are some differences concerning the question whether responsibility has to be proved and who has the burden of doing so or proving the contrary. **Germany** is more strict in this respect than **Denmark**, **France** or the **Netherlands**, although in the latter country there is a tendency of the courts to transfer the burden to the defendant. That the licensor is obliged to assist or to guarantee technical performance to some degree is implied by the laws of Denmark, Germany, **Greece**, **Italy**, **Japan**, **Korea**, **Spain**, **Switzerland** and **Turkey**, save there is another provision in the agreement. In any case at the moment, **Germany** seems to be the pioneer in following the footsteps of the Americans.

One of the most spectacular cases in Germany was when a motorcycle manufacturer was held liable because of a motorcycle onto which the owner had fixed a component produced by another manufacturer, and the rider driving excessively fast was killed. The reason why the Court came to this verdict was that the motorcycle manufacturer had, in fact, provided some holes to enable the mounting of this component which had led to the death of the user. No doubt, such decisions put more risks onto sellers, licensors and their contracts, such as also license agreements. The result is that more precaution should be exercised.

Precautionary measures
There are three groups of remedies to minimize those hazards and these should be carefully considered.

One is a good insurance and it is worth checking up whether cases of tortious liability are covered within the framework of your ordinary professional liability insurance.

Secondly, care should be taken when supplying parts, e.g. within the framework of a Know-How license. You have to stretch your imagination when considering how your product could be used - or abused. Then it may become necessary to revise your manufacturing practice, to instruct your internal quality control and to take care to ensure a precise quality documentation in order to be able to prove that any failure is not your responsibility. You should also ensure that instructions for storing, transporting and use are appropriate so that nobody comes to harm. And you should even consider the case of a modification of the product by a third party or the incorporation or assembly with other parts.

The legal remedy is to try to exclude the risks by an appropriate formulation of license contracts. It is clear from the foregoing explanation that this cannot be done with a particular format, because of the different types of licenses, the different risks, the countries involved and the question whether the contract has to be established for the licensee or the licensor.

Therefore, I can give you only some suggestions and examples. In the case of a trademark license, I have already stated that the licensor will ordinarily be made liable. This is due to the fact that the consumer will be under the impression that he has manufactured the hazardous product. In order to avoid this a clause might be used that the licensee is obliged to put an inscription "Made by ..." onto the product. A simple notice "Made in ..." would not be sufficient, it must be the name. In other license agreements, it might be possible to restrict liability by contract in various respects and ways which has to be tailormade in every case. Even a clause that the agreement will be terminated if the product runs into difficulty can be a way of avoiding liability. Or in cases where the licensee is entitled to print the licensor's name ("Licensed by ...") one should insert a caveat that the name should not appear as a trademark.

In other contracts one should have a rule for the liability in case of an infringement into rights of third parties. Depending on the economic and other potential, the liability should be transferred to the stronger party or should be limited, if possible, by agreement for that party which is in the weaker position. And both parties can indemnify themselves against the possibility of a third party having their rights infringed.

A special topic is also Know-How licenses where the licensor will be liable when the transferred Know-How does not produce the performance stated in the contract; he should not be made liable in case of errors or lack of information by the licensee; he can avoid liability, if the licensee does not begin to work within a reasonable time or the correct time. Liability in some cases, such as force majeur, for example a war or a catastrophe, can also be excluded or when the licensee supplies some of his own equipment incorporating the licensed product. The licensor or the licensee can even limit his liability up to a certain maximum.

In case of a mixed license, it will be necessary to demarcate the rights and duties for one type of license and the other, so that discussions can be avoided in case of a later change of relationship between the types of licenses contained in the agreement, such as the protective right or one of them turning out to be invalid as a whole or in part.

To conclude, I wanted to show you with my lecture that, even if you master the problem of coming to an agreement with your partner, there remains a lot to do to ensure that other problems will not arise later on.

10 FORENSIC ENGINEERING ACTIVITIES IN THE CZECH REPUBLIC

MILOŠ DRDÁCKÝ

Academy of Sciences of the Czech Republic, Prague, Czech Republic

INTRODUCTION

In the Czech Republic the expert opinions are mostly elaborated by so called experts or forensic engineers. Persons or institutions which are entitled to act as expert witness are established by the Ministry of Justice or by chairpersons of regional courts. The experts are listed and published according to the 49 branches summarized in the Appendix. There exists a possibility to invite unlisted top specialists for solutions of special questions in individual cases. All procedures and necessities are given in two laws - the law No. 36/1967 on forensic experts and interpreters and the executive law No. 37/1967 with two recent amendments.

The listed experts are approved to submit expert opinions in connection with litigations or generally at any legal acts of persons or other subjects. In other words their participation is necessary at any proceeding which combines law and technical approaches.

Possible candidates for acceptance as forensic experts have to fulfill four personal or professional conditions. They must be of the Czech citizenship - this condition can be exceptionally excused; they must have necessary knowledge and experience in the relevant field, preferably supported by a special educational background, if it is available in this field; they must have adequate personal character and they must agree with the nomination. We have at present about 10 400 listed experts, about 4 050 of them being appraisers.

EDUCATIONAL POSSIBILITIES

Professionals interested in forensic engineering activities have several possibilities to improve their basic educational background. Nevertheless, we have no special schools or full study programs in this field. There exist specialized courses for both undergraduate as well as graduate professionals, moreover some universities have in their student programs courses of forensic problems, e.g. the Technical University in Brno has a specialized Institute of Forensic Engineering.

Of course, the branches and specializations of the above mentioned courses are dependent on market demands for expert opinions. Here the majority of experts in the Czech Republic is focused on problems of valuation of fixed assets (real estate) and car assessments, which can be carried out even by non-graduate professionals.

Structural Failure: Technical, Legal and Insurance Aspects. Edited by H.P. Rossmanith. Published 1996 by E & FN Spon, 2–6 Boundary Row, London SE1 8HN, UK. ISBN: 0 419 20710 4.

PROFESSIONAL ACTIVITIES OF FORENSIC ENGINEERS

Professional bodies

The experts have created several professional associations. The most important body under the name "Chamber of Forensic Experts of the Czech Republic" assembles about 2 200 experts who work in five divisions:

- civil engineering including the real estate assessors,
- transport and machinery including car assessors,
- electronics,
- medical and psychological experts,
- other specialized assessors and experts.

Similar organization has been created under the name "Association of Experts and Assessors of the Czech Republic" and it joints about 200 persons.

"The Czech Chamber of Appraisers", a member of TEGOVOFA, associates very specialized listed or unlisted experts which are mostly engaged in valuation problems of real estate for finance industry.

Some other minor specialized groups of listed or unlisted experts created various organizations, as e.g. "The Association of Non-Destructive Diagnostics Engineers".

The Czech forensic experts feel the lack of a legislative professional body which should govern the profession, promote educational activities, maintain ethics of experts, perform examinations and give recommendations to approve or revoke licenses, etc. Therefor, we have prepared a legislative proposal for a Chamber of Forensic Engineers established by a law.

Conferences

Since 1991 an International Conference *Lessons from Structural Failures* has been organized and the presented papers are published in Proceedings of the same title. This year the third conference was announced as Regional Symposium of ISTLI. The call for papers for the fourth conference are available at the symposium desk. The conference will be concentrated on non-bearing building structures and furniture.

FORENSIC ENGINEERING PROBLEMS IN THE PERIOD OF THE CZECH ECONOMY CONVERSION

The present state in the Czech Republic is characterized by a massive change of laws, which brings about some difficulties and does not enable an easy orientation in the legal environment. For example, our system of building, urban planning and heritage preservation laws is going to be changed very substantially.

Many new problems arose from the restitution laws. They cover not only technical problems but also medical. In the technical fields, there are connected mostly with litigations about differences of the character, state or value of real estate before communist seizure and at present.

Another large family of cases occurred due to very non-qualified, non-professional or incompetent approaches to the ever growing business and undertaking activities, especially in civil engineering. A large demand for shops, restaurants, exhibition halls, etc. lead to many reconstructions or changes carried out without a proper documentation, without sufficient experience and even by non-qualified people. There are many cases of total failures only in Prague. Some new constructional firms have not appropriate technological knowledge of modern and foreign techniques.

There was noticed an increase of litigations concerned with environmental problems or conflicts between neighbours, as e.g. shading of the land.

THE CZECH NATIONAL GROUP OF ISTLI

The Czech forensic experts and their activity can be improved by an international exchange of knowledge and experience between forensic engineers. Therefor, the Czech experts welcome the establishment of an international society in this field and some of the created a nucleus of the Czech National Group which intends to be incorporated into the ISTLI. We have prepared a draft of Statutes which were widely discussed and we expect the approval of our Group by the Ministry of Interior at the beginning of the year 1994.

11 SOME COMMENTS ON RISK ASSESSMENT AND FAILURE ANALYSIS ACTIVITIES IN SWITZERLAND

R. KIESELBACH
EMPA-Dübendorf, Dübendorf, Switzerland

Summary

Since the catastrophic fire of Sandoz company in Basel/Switzerland in 1986 legislation in Switzerland in the field of hazardous materials has been improved. This led to mandatory risk assessment of major industrial hazards.
Because support of these investigations by science is needed a major research project "Safety and Risks of Technological Systems" was initiated by the Swiss Federal Institute of Technology (ETH) in Zürich/Switzerland.
Scope and preliminary results of this project are discussed in this paper.

A second topic of this paper are the activities in the field of case studies and failure analysis of buildings and components in Switzerland.
The Swiss Federal Laboratories for Materials Testing and Research (EMPA) have a special branch for "Building Components/Damages" to investigate failure and damages of buildings.
Failure analysis of metallic components and structures has been reorganized and offers improved services to the public and for research.
This is briefly discussed in the paper.

Risk and Safety of Technological Systems

In 1986 a warehouse with chemicals of Sandoz company in Basel/Switzerland burnt down. The developping fumes threatened the population nearby because their composition was unknown. The explosion hazard could not be assessed because the information on the exact contents of the warehouse was not immediately available. Furthermore there was no containment for the water contaminated after use by the fire brigade so that the river Rhein was severely polluted.
This major incident led to the introduction of the Swiss Major Hazard Ordinance: "Störfallverordnung", similar to the corresponding german law, which makes risk assessment mandatory for major risks like production, storage and transportation of hazardous substances.
This Swiss regulation became effective without sufficient theoretical basic information how to perform risk assessment with respect to national conditions and without sufficient national data on amount and location of production, storage and transport of hazardous substances.

Structural Failure: Technical, Legal and Insurance Aspects. Edited by H.P. Rossmanith. Published 1996 by E & FN Spon, 2-6 Boundary Row, London SE1 8HN, UK. ISBN: 0 419 20710 4.

Around 1988 members of the professorate of the Swiss Federal Institute of Technology (ETH) in Zürich, the so called Polytechnikum, felt that they ought to contribute to improving the situation and thus in 1991 the socalled "Poly-Projekt" was started.
Its full title is "Risk and Safety of Technological Systems"
Most important object is a description of the state of the art of risk analysis, risk assessment and risk management and the development of practical methods for establishing a regional safety plan. This safety plan should not only consider technical problems but make also use of the multidisciplinary facilities of an institution like ETH.

Additional goals were the improvement of cooperation between different authorities in national administation and industry and assistance to tuition e.g. by post-graduate courses in risk assessment etc.

The project started in 1991 and the first phase will end 1994. The annual funds by ETH are 500'000.- Swiss Francs; additional funds are granted by the cooperating institutions.

At the end of 1992 a first status report was issued after a three day seminar on the results achieved until then which took place in Ascona/TI in Switzerland.

A second three day seminar in Ascona has taken place on Nov. 1, 1993 and was focused on the topic of the contributions already available for the handbook on regional safety management in Switzerland.

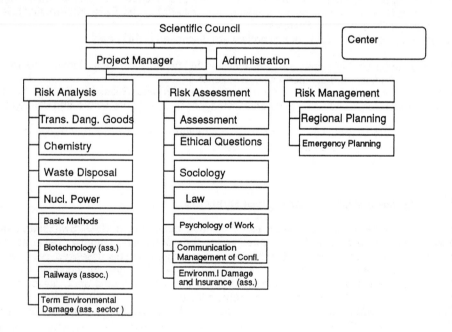

The project is organized as shown.
This makes it clear that a lot more than only technical aspects are covered by this project.
Final goal of the project, apart from the handbook for a regional safety plan, is to establish a new chair for Risk Science at ETH in Zürich or Lausanne.

ublished results at present are:

Gesellschaft-Ethik-Risiko
Resultate des Polyprojekt-Workshops vom 23.-25.11.1992, Ascona
Ed. H. Ruh/H.Seiler
Verlag Birkhäuser, Basel 1993

Operationelle Routenwahl im Gefahrenguttransport
G.E.G. Beroggi et al.
Polyprojekt-Dokument Nr. 1
Verlag der Fachvereine, Zürich 1993Report of Ascona Seminar 1992

Sicherheit in soziotechnischen Systemen
G. Grote/C.Künzler
Polyprojekt-Bericht Nr. 5/93
Zentralstelle ETH-Polyprojekt, Zürich Sept. 1993

Raumplanung und Störfallrisiken
C.S.Furter/R.Simoni
Polyprojekt-Bericht Nr. 6/93
Zentralstelle ETH-Polyprojekt, Zürich Juli 1993

Was ist regionale Sicherheitsplanung
Resultate Polyprojekt-Workshop 10./11.5.1993
Polyprojekt-Bericht Nr. 7/93
Zentralstelle ETH-Polyprojekt, Zürich Aug. 1993

raft:
Integrated Regional Risk Assessment and Safety Management
A Guideline for Systems Engineers and Decision Makers
Ascona, Nov. 1993

Failure Analysis Activities at the Swiss Federal Institute for Materials Testing and Research (EMPA)

EMPA is the national laboratory of Switzerland for Materials Testing and Research. It has 3 major divisions and more than 40 sections which work in different fields of engineering and science.Its purpose is to render services to the public in various technical fields and o assist the Federal Institute of Technology and to help the Federal administration in the application of legislation.
EMPA has 2 main things to offer:
1. It is acknowledged for its competence in specialized fields of technology.
2. It is able to perform routine services very efficiently.

This simplified graph shows the main activities of EMPA concerned with failure analysis in ne way or the other.

n 3 sectors we have a special organization to deal with failure analysis. Contracts with companies and private persons are accepted on the condition that the order is signed by both parties. Furthermore expert evidence for courts of law is given frequently.

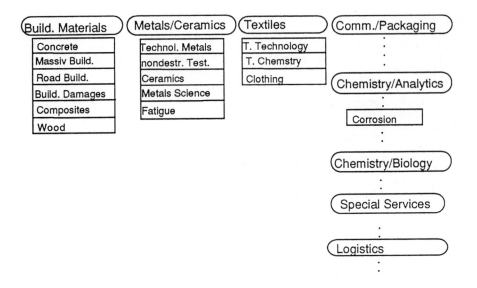

Building Damages

This sections main field of activity is:

recording of major damages in buildings, documentation
establishment of the cause for the damages
consulting for preventive measures
research
giving expert evidence

Most important for the work of this section is the collection of case studies and files with more than 2'500 diferent topics and more than 20'000 colour slides.
This documentation is at present being stored in a computer database to facilitate access to the information. Main piece of work is apart from scanning of the records including drawings and photographies the choice of the proper clues for each file.

Examples: *collapse of buildings*
cracking of walls or plaster
leaking of flat roofs

These records are not open to the public. Therefore regularly reports are issued on important or frequent cases of failure or damage and seminars are organized on special topics.

Damage of Textiles

The division of EMPA in St. Gallen has gained a vast expert knowledge on damages or quality problems with textiles. This knowledge is collected now in a computerized expert system.
The collected knowledge is structured and put to use by definition of certain rules connecting objects and records, characterizing objects by certain properties and giving appropriate boundary conditions.
The expert system is based on the software package "Diagnostic Master".
The user who must possess at least basic knowledge of textile technology is guided through a menu by answering questions on the problem and the object at hand. After working through this tree of decisions he finally gets to the solution of his problem.

Examples :   *strips or knots in a fabric*
                        waviness of a fabric

Failure of metallic components

Because of the importance of metals as materials a whole division of EMPA is concerned with their different aspects:
  technology of metals/joining
  metals science/surface technology
  metals/ceramics
  non destructive testing
  dynamic testing and stress analysis

In addition experts from other sections, e.g. in chemistry, corrosion and measurement can assist the case studies.
Responsibility for project management is with a staff-position which also installs the necessary ad hoc organization for the individual projects.
Cases with a lesser interdisciplinary character are handled by the section which is mainly concerned.
This reorganisation has had the aim to deal with case studies in a more systematic way instead of the former intuitive approach. For this purpose checklists are used and a computer-software is being developped, where the different laboratories or sections or groups working together on a case collect their information and their results in a common database which is accessible by the shared computer network.

Examples: *bursting of a chemical reactor for ammonia*
                        failure of a cable car
                        successive failure of several wire ropes of a long suspension bridge

12 THE FUNCTION OF TECHNICAL DOCUMENTATION: DRAFT RECOMMMENDATION VDI 4500 FOR USER INFORMATION

C. O. BAUER
HDI, Wuppertal, Germany

Instruction Responsibility as Vital Part of Product Liability

Within the frame work of product liability responsibility for users information is one of the important parts. Picture 1 defines these requirements, product unspecific, according to general legal demands.

It is the responsibility of the manufacturer - or importer - to give the foreseeable user all necessary information to use the products in its various applications without danger or avoidable risks.

The legal addressee of these informations is the foreseeable user with his average knowledge and abilities, how difficult it may be to identify these abilities of users groups and their various qualifications.

If by appropriate understandable information a misuse or damage could have been prevented, defects in users instructions are as important as any technical defects in construction or during production.

As this is a complicated intersection between manufacturers experiences, knowledge and those of completely different acting users, to fulfil this responsibility includes psychological, as well as intellectual and administrative problems at large.

The German High Court decided a number of cases, but a complete system of all legal demands has not yet been systematically established neither anywhere in Europe nor in the USA. All requirements have to be differentiated according to the various products — product specific — and their different application.

Structural Failure: Technical, Legal and Insurance Aspects. Edited by H.P. Rossmanith. Published 1996 by E & FN Spon, 2–6 Boundary Row, London SE1 8HN, UK. ISBN: 0 419 20710 4.

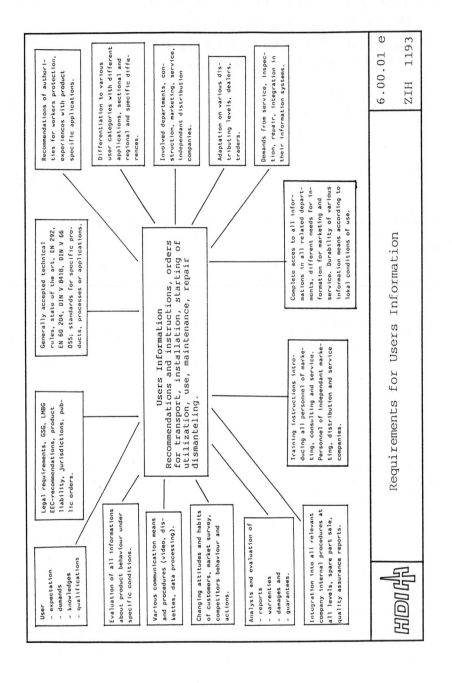

User
- expectation
- demands
- knowledges
- qualifications

Legal requirements, GSG, LMBG EEC-recommendations, product liability, jurisdictions, public orders.

Generally accepted technical rules, state of the art, EN 292, EN 60 204, DIN V 841B, DIN V 66 055; standards for specific products, processes or applications.

Recommendations of authorities for workers protection, experiences with product specific applications.

Differentiation to various user categories with different applications, sectional and regional and specific differences.

Involved departments, construction, marketing, service, independant distribution companies.

Adaptation on various distributing levels, dealers, traders.

Demands from service, inspection, repair, integration in their information systems.

Evaluation of all informations about product behaviour under specific conditions.

Various communication means and procedures (video, diskettes, data processing).

Changing attitudes and habits of customers, market survey, competitors behaviour and actions.

Analysis and evaluation of
- reports
- warrenties
- damages and
- guarantees.

Integration into all relevant company internal procedures at all levels, spare part sale, quality assurance reports.

Users Information
Recommendations and instructions, orders for transport, installation, starting of utilization, use, maintenance, repair dismanteling.

Training instructions introducing all personnel of marketing, consulting and service. Personnel of independant marketing, distribution and service companies.

Complete acces to all informations in all related departments, different needs for information for marketing and service. Durability of various information means according to lokal conditions of use.

Requirements for Users Information

6.00.01 e

ZIH 1193

HDI

EEC Recommendation Machines as Guideline

With the 1st January 1993 beginning of the free european market —
transaction period ending at the 31st December 1994 — the mandatory
requirements of the EEC Recommendation Machines has established a
minimum level of criteria, which producers — and importers — have
to observe. Picture 2 and 3 give the concentrated content.

These minimum requirements are not sufficient to fulfill in respect
to the overall legal demands on product liability. These guidlines
have to be filled out product specific and according to the various
applications of the different products. According to the text of the
recommendation, they are minimum — but not sufficient — to conform
with legal requirements. But still these minimum requirements are
not yet overall practice for many european manufacturers and importers of third country products. It has to be recalled that from 29th
June 1994 the EEC Recommendation about general safety for consumer
products will be valid additionally. The legal demands for all products intended to be used by private consumers have then to reach the
same safety level as machines. The EEC Recommendations do not contain specific and detailed technical figures to observe. The EEC
Recommendation demands in its article 3 very simple:

"The manufacturer has to supply only safe products."

Giving no detailed values how and in what way this should be accom-
plished.

The Important Role of Technical Documentation

The various kinds of Conformity Certificates for industrial products
for which EEC Recommentations exist — see picture 4 — allow to use
different models. But all models demand complete and reliable technical documentation, composed out of

 * internal documentation from
   design, construction, production, assembling and test procedures with reliable technical datas and
 * complete and understandable users information
   aimed at the average foreseeable user with his (minimum)
   knowledges and expected abilities.

Instructions

(a) All machinery must be accompanied by instructions including at least the following:

— a repeat of the information with which the machinery is marked (see 1.7.3), together with any appropriate additional information to facilitate maintenance (e.g. addresses of the importer, repairers, etc.),

— foreseen use of the machinery within the meaning of 1.1.2 (c),

— workstation(s) likely to be occupied by operators,

— instructions for safe:

— putting into service,

— use,

— handling, giving the mass of the machinery and its various parts where they are regularly to be transported separately,

— assembly, dismantling,

— adjustment,

— maintenance (servicing and repair),

— where necessary, training instructions.

Where necessary, the instructions should draw attention to ways in which the machinery should not be used.

(b) The instructions must be drawn up by the manufacturer or his authorized representative established in the Community in one of the languages of the country in which the machinery is to be used and should preferably be accompanied by the same instructions drawn up in another Community language, such as that of the country in which the manufacturer or his authorized representative is established. By way of derogation from this requirement, the maintenance instructions for use by the specialized personnel frequently employed by the manufacturer or his authorized representative may be drawn up in only one of the official Community languages.

(c) The instructions must contain the drawings and diagrams necessary for putting into service, maintenance, inspection, checking of correct operation and, where appropriate, repair of the machinery, and all useful instructions in particular with regard to safety.

Instructions for Machinery
Minimum Content I
§ 1.7.4, Directive 891392 EEC

(d) Any sales literature describing the machinery must not contradict the instructions as regards safety aspects; it must give information regarding the airborne noise emissions referred to in (f) and, in the case of hand-held and/or hand-guided machinery, information regarding vibration as referred to in 2.2.

(e) Where necessary, the instructions must give the requirements relating to installation and assembly for reducing noise or vibration (e.g. use of dampers, type and mass of foundation block, etc.).

(f) The instructions must give the following information concerning airborne noise emissions by the machinery, either the actual value or a value established on the basis of measurements made on identical machinery:

— equivalent continuous A-weighted sound pressure level at workstations, where this exceeds 70 dB(A); where this level does not exceed 70 dB(A), this fact must be indicated,

— peak C-weighted instantaneous sound pressure value at workstations, where this exceeds 63 Pa (130 dB in relation to 20 µPa),

— sound power level emitted by the machinery where the equivalent continuous A-weighted sound pressure level at workstations exceeds 85 dB(A).

In the case of very large machinery, instead of the sound power level, the equivalent continuous sound pressure levels at specified positions around the machinery may be indicated.

Sound levels must be measured using the most appropriate method for the machinery.

The manufacturer must indicate the operating conditions of the machinery during measurement and what methods have been used for the measurement.

Where the workstation(s) are undefined or cannot be defined, sound pressure levels must be measured at a distance of 1 metre from the surface of the machinery and at height of 1,60 metres from the floor or access platform. The position and value of the maximum sound pressure must be indicated.

(g) If the manufacturer foresees that the machinery will be used in a potentially explosive atmosphere, the instructions must give all the necessary information.

(h) In the case of machinery which may also be intended for use by non-professional operators, the wording and layout of the instructions for use, whilst respecting the other essential requirements mentioned above, must take into account the level of general education and acumen that can reasonably be expected from such operators.

Instructions for Machinery
Minimum Content II
§ 1.7.4, Directive 891392 EEC

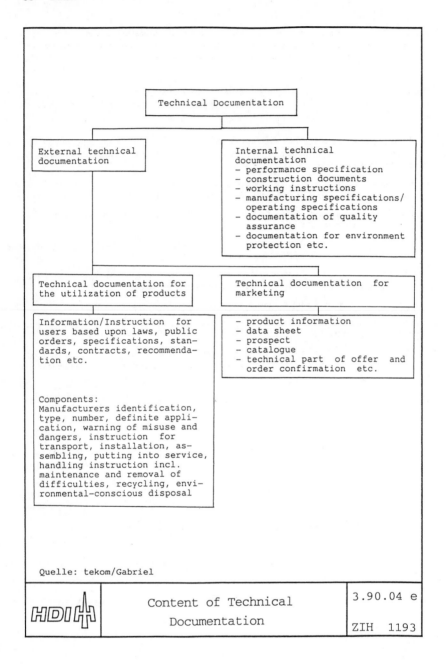

Technical Documentation

External technical documentation

Internal technical documentation
- performance specification
- construction documents
- working instructions
- manufacturing specifications/ operating specifications
- documentation of quality assurance
- documentation for environment protection etc.

Technical documentation for the utilization of products

Technical documentation for marketing

Information/Instruction for users based upon laws, public orders, specifications, standards, contracts, recommendation etc.

- product information
- data sheet
- prospect
- catalogue
- technical part of offer and order confirmation etc.

Components:
Manufacturers identification, type, number, definite application, warning of misuse and dangers, instruction for transport, installation, assembling, putting into service, handling instruction incl. maintenance and removal of difficulties, recycling, environmental-conscious disposal

Quelle: tekom/Gabriel

| HDI | Content of Technical Documentation | 3.90.04 e |
| | | ZIH   1193 |

For all models the total technical documentation has to be presented to state authorities in case of damages or control. For other models it will be checked by "notified organization units" according as part of design tests. The content and explicitivity of users information has been many times topic for american court decisions and constitute as well technical as administrative and psychological problems, to be filled out in the specific cases the over all general legal terms sufficiently detailed as to the assumptions of the various courts.

Draft Recommendation VDI 4500 Users Information

<u>Systematical Approach</u>

This highly complicated and often dangerous situation for the manufacturers and the lack of any reliable and detailed standards have caused the German Association of Engineers (Verein Deutscher Ingenieure) to establish in 1990 a committee Technical Documentation. Members came from producing and distributing companies, universities, free lance agencies, insurance companies and practising technical writers. As far as possible all groups were represented with interests in and obligations to the various fields and sections of technical documentation.

After intensive committee work it presented just now the draft of VDI 4500 concentrating in its content on problems and solutions of users informations. Following second and third parts will describe the problems of internal technical documentation where standards with specific solutions already do exist.

This draft standard is published with the intention of qualified thorough testing in day-to-day work of companies of all sizes and kinds to establish and prove themselve as necessary tool to fulfill overall general legal demands with usable technical solutions. Any proposals for alteration and amendment are welcome so that it might be completed in due course as generally accepted technical rule according to the legal definition and the requirements of EEC technical law.

Extensive Content

Picture 5 shows the content. It starts with a short explanation of
the different legal requirements as the limits which are set for
users informations from product liability as well as by contractual
law.

It is followed by the principles of formation with wording, construction and – one of the most decisive elements – of appropriate
language for the understanding and the knowledge of the different
users groups.

The organisational requirements to establish understandable users
informations is a wide field, systematically not really organized by
companies of various sizes, products and for varying user groups.
The organisational necessities are described as part of a steering
process in marketing, sales and service. This is part of the freedom
of organization of the companies to form their internal organization, but at the same time the companies must be able to prove that
they have evaluated to the necessary extent all unspecific legal demands as to reasonable caution and reliability of informations and
evaluation of product observation upon short notice.

The users service is one of the most decisive element as mandatory
part of users information. Human interface design is therefore explained with its necessities as part of users information, to inte-
grate a new and often overseen modern tool.

Modern technique to establish and to produce users informations is
in a rapid change; it is therefore impossible to give an overall
summary of all available techniques with their hard- and software at
the moment of printing or even for a short period thereafter. General characteristics for selection and use therefore constitute the
frame work, in which the companies have to make their own selection
as to their requirements and products of available hardware.

Unrealistic assumptions about costs are the reason for much too many
insufficient and unacceptable solutions for user information in many
companies. Calculation schemes and principles for economical control
are given, to help the companies to find the individual best solutions, which fulfills the legal demands with economical justifiable
costs.

Technical Documentation

Part 1 User's Information

Technical Documentation \ Department	Construction L	Construction S	Production L	Production S	Quality Assurance L	Quality Assurance S	Purchasing L	Purchasing S	Marketing L	Marketing S	Service L	Service S	Indep. Distrib. Companies	...	Management
Product Presentation															
– advertising	M				I				(V)		M		I		M
– product information	(V)								M		M		M		M
– catalogue			M			I			(V)			M	I		I
– data sheet		(V)				I				M		M	I		I
– users information	M						I		(V)		M		M		M
– application advices	M				M				(V)		M		M		I
Product application															
– packing	M	M			M		M		(V)		M		M		I
– marking	(V)	M			M			I		M	I		I		I
– users instruction	M	I			I					M	(V)		M		I
– users training	M							I	M		(V)		M		I
– service	I					I	I				(V)		M		I
* maintenance															
* repair															
– Recycling	M			I	I					M	(V)		M		M
Product Oberservation															
– collection of field experiences	M				I				M		(V)		M		I
– treatment of complaints	M		M		M				(V)		M		I		I
– damage analysis	M		M		M				M		M		I		(V)
– evaluation of	M		M		M				M		M		M		(V)
* users information															
* limitation of application															
* exchange															
* replacement															
* recall															

L = Leader, department head   S = Specialist

M = Participation   I = Information   V = Responsibility

Responsibility and Participation of Different Departments for Tasks and Processes of Technical Documentation	3.00.16 e	
	ZIH	1193

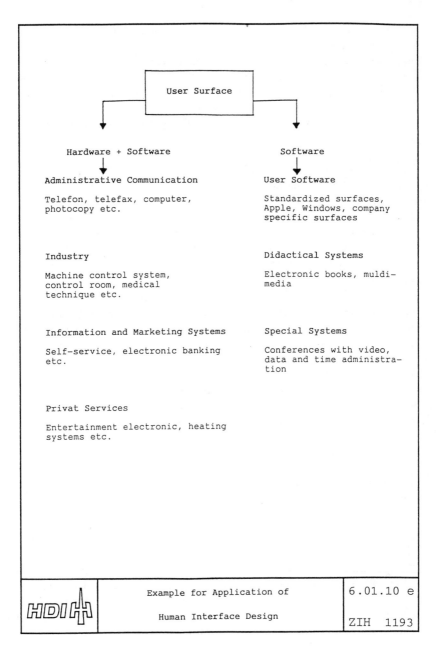

Example for Application of	6.01.10 e
Human Interface Design	ZIH 1193

How to organize the process of preparing users informations are in many companies an example for trial and error, just resulting from varying chances rather than by rational constructed administration. The state of the art as to organisational series and practical necessities are presented as guidelines to which company internal practices could be evaluated by management as well as by external "notified organisational units", if EEC-conformance certificates have to be issued.

Long abstract thesis and explanations are in most cases of no practical help for the companies. Therefore the draft standard contains a selection of checklists for the content, the process of composition and all those criteria which should be observed during planning, production and testing users information. Checklists alone are of no help as long as they are not systematically constructed and take into account all the various aspects of the users information. Opinions alone might be of interest in social discussions, but they are not helpfull in discussions about workable organisational setups. Only if their basic knowledges are solid and as extensive as possible to constitute a sound fundament for solutions, which will stand up if they might be tested by experts or in courts it has reached reliable management targets.

Clear definitions of necessary concepts are the only base for clear and rational thinking as a durable fundament for economical justifiable actions. Evaluating legal definitions, technical processes and organisational set-ups, a collection of the necessary concept try to create a sound and - hopefully - non-disputable base for further discussions and evaluations on this very important difficult borderline between various internal departments and external requirements and expectations.

Evaluation in Other Regions

This draft standard was composed and drawn up regarding only the very specific conditions and demands in Germany. It did not try in this first stage to include the differences which might occure addapting and applying it to other regions or states in or outside the EEC as it is a mandatory request of the EEC Recommendation Machine too.

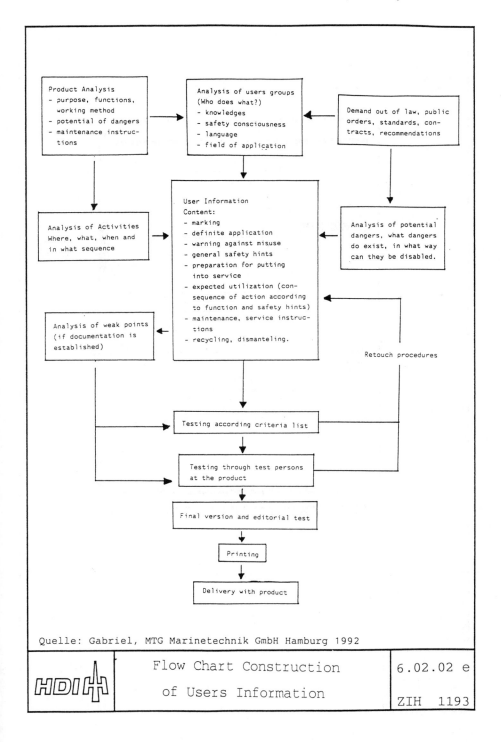

Product Analysis
- purpose, functions, working method
- potential of dangers
- maintenance instructions

Analysis of users groups (Who does what?)
- knowledges
- safety consciousness
- language
- field of application

Demand out of law, public orders, standards, contracts, recommendations

User Information
Content:
- marking
- definite application
- warning against misuse
- general safety hints
- preparation for putting into service
- expected utilization (consequence of action according to function and safety hints)
- maintenance, service instructions
- recycling, dismanteling.

Analysis of Activities
Where, what, when and in what sequence

Analysis of potential dangers, what dangers do exist, in what way can they be disabled.

Analysis of weak points (if documentation is established)

Retouch procedures

Testing according criteria list

Testing through test persons at the product

Final version and editorial test

Printing

Delivery with product

Quelle: Gabriel, MTG Marinetechnik GmbH Hamburg 1992

	Flow Chart Construction	6.02.02 e
HDI	of Users Information	ZIH 1193

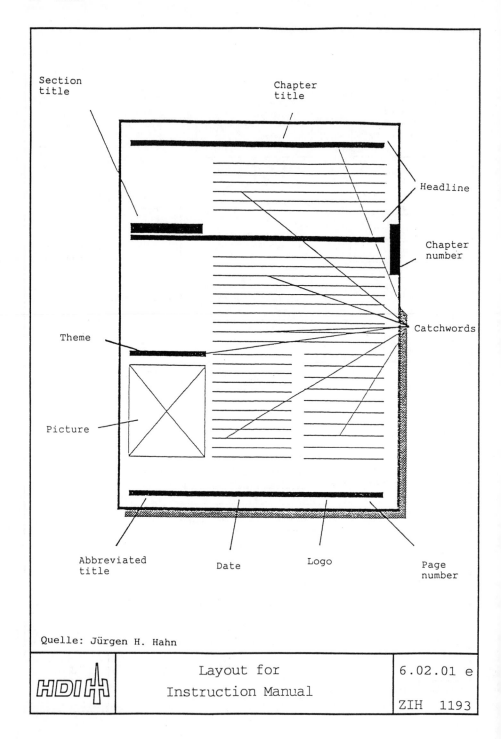

Many of the legal demands and requirements are universe and constitute general accepted conditions. Others may differ in their limits and borderlines. For the conditions of other regions it is possible and foreseeable that additional conditions and aspects have to be added and observed as to regional specific criteria. Human attitudes, expectations and human behaviour in companies as well as by user groups might constitute different approaches and additional necessities.

To initiate discussions about positive and workable solutions this draft recommendation is presented to experts in other countries causing their interest and hoping of their discussing part.

To compete with the legal demands and to fulfill users expectations for safe products comming up to the intended value under the various different applications is a universal task. The result of such discussions should be formulation, a definition of means and ways to give sound solutions for these universal questions and tasks – with the necessary diferentiation to the various products, local different attitudes with necessary technical substance.

13 INTERNATIONAL COMMERCIAL ARBITRATION AND ENGINEERING

F. SCHWANK
Attorney at Law, Vienna, Austria

Arbitration is perhaps best understood to be a private, that is a non-state, court system of dispute resolution. A dispute, instead of being filed at a state court and decided by a state judge, is referred to a private judge, an arbitrator or an arbitral tribunal, either directly or through an arbitration institution.

Arbitration will be discussed in the context of international engineering disputes. Generally, these disputes arise between parties from different countries and concern engineering or related agreement(s), provision or lack thereof of engineering services, the supply of plants and technology, building contracts, or the supply of utility facilities.

Within this context, the ensuing discussion summarises the differences between arbitration and litigation, describes the various types of arbitration and the relevant arbitration bodies, and discusses the advantages and disadvantages of conducting arbitration proceedings. Finally, I will comment on the drafting of different aspects of the arbitration clause.

INTERNATIONAL ARBITRATION OR LITIGATION

Litigation is the resolution of a dispute through the State courts whereas arbitration is resolution by an arbitral tribunal. While the State court will always be a national institution, an arbitral tribunal for an international arbitration will usually be composed of arbitrators of different nationalities.

Structural Failure: Technical, Legal and Insurance Aspects. Edited by H.P. Rossmanith. Published 1996 by E & FN Spon, 2–6 Boundary Row, London SE1 8HN, UK. ISBN: 0 419 20710 4.

There are several important characteristics of arbitration which distinguish it from litigation. The most important of these are the following:

1. There must be an agreement between the parties to use arbitration.

The agreement may be incorporated into the contract between the parties themselves or may be made by the parties at a later time, e.g., when a dispute arises – although at that time it will be more difficult to obtain an agreement between the parties. The agreement must be in writing, but need not be a formal written agreement. An exchange of letters, faxes, telexes or EDI messages will suffice.

2. Arbitration is private

The arbitral process is strictly private. Details of the proceedings and any award made are not published unless with the agreement of the parties. No other person is entitled to become involved. Only the parties to the arbitration agreement, their lawyers, the witnesses and any experts appointed may participate. No other party may or can be forced to be a party to the proceedings.

Contrast this to the practise in many countries, in particular the common law countries which are based on the English system, where, once a claim has been filed at the court, it becomes a public matter which may be reported in the newspapers and any person may attend hearings, unless specifically prohibited by the judge by way of a reasoned decision. It is also possible to join other parties to the proceedings as a "third party" or for other persons to "intervene" in the proceedings.

3. The parties have control over the arbitral process

The parties have considerable control over the way the arbitration is conducted because they can themselves determine its procedural rules. For example, they can decide:

- how many arbitrators there will be;
- who will be the arbitrators;

- who will be appointed as technical expert(s);
- whether there will be oral hearings or only written pleadings and supporting documents submitted;
- in which language or languages the arbitration will be conducted;
- where the arbitration will take place.

In litigation, the conduct of the proceedings is, to a large extent, out of the control of the parties. Once a claim is filed at a court, the court will then determine how and when the matter will proceed. At best, the parties or their counsel can make submissions or proposals, but it is the judge who will always ultimately decide when and how the matter will be heard.

TYPES OF ARBITRATION

There is quite a large range of options available for parties who have agreed or are considering the option of using a different method of resolving disputes.

1.   Institutional arbitration

The parties can have a dispute resolved by institutional arbitration. If they do so, the parties then agree that the arbitration will be conducted in accordance with the arbitration rules of a permanent supervisory or administrative institution. There are many such institutions throughout the world and the one chosen by the parties will depend on such factors as the industry or branch of business of the parties, the place of the performance of the contract between the parties, the purpose of the contract, the experience of the parties, and the expertise and location of the arbitral body.

Some of the relevant institutions are:

a) The International Chamber of Commerce (ICC). This institution has been providing arbitration facilities since 1923. The Rules of Conciliation and Arbitration contain detailed provisions on how the arbitration will be conducted and provide for close supervision of an arbitration by the International Court of Arbitration of the ICC.

b) London Court of International Arbitration.

c) American Arbitration Association.

d) International Centre for the Settlement of Investment Disputes (ICSID).

e) Chambers of Commerce. For example, in Austria, the Arbitral Centre of the Federal Economic Chamber (Bundeskammer der Gewerblichen Wirtschaft) has international arbitration rules and facilities for disputes when at least one party is a non-Austrian.

f) Commercial and trade associations. These associations often provide arbitration rules for disputes relating to their particular trade or activity or if at least one party is a member of the association.

Each of these institutions has a set of rules to be applied in any arbitration. The institution will then administer the arbitration and supervise the conduct of the arbitration, in particular, ensuring that it is conducted in accordance with the rules. The amount of supervision varies, with the most being from the ICC which requires copies of all pleadings and correspondence and scrutinises the award before it is released to the parties.

A party to a dispute will apply to the institution with the request that his claim be settled by arbitration. The institution will then set into motion the arbitration process, including the procedure for appointment of the arbitrator or the arbitral tribunal agreed by the parties or provided for in its arbitration rules.

2. Ad hoc arbitration

If the parties choose ad hoc arbitration, the parties themselves decide on the appointment of the arbitrators and on the rules to be applied to an arbitration. The rules may be those agreed by the parties or may be the arbitration rules of a body such as the United Nations Commission on International Trade Law (UNCITRAL).

These ad hoc arbitration rules may be amended by the parties and in fact the rules state something like: "Unless otherwise agreed by the parties...."

There is no supervision or scrutiny of the arbitration by any institution other than the arbitrators themselves. Most administrative duties are carried out by the arbitrators.

When the parties cannot agree on the appointment of an arbitrator, the problem can be referred either to the State court or to a designated person for resolution. Such a person could be the president of the local chamber of commerce where the arbitration is to be conducted. The UNCITRAL Rules also provide that the Secretary-General of the Permanent Court of Arbitration at The Hague shall designate the appointing authority for ad hoc arbitrators.

3.   Alternative Dispute Resolution

Instead of using these more formal and litigious procedures for resolving disputes, it is also possible for the parties to use a less formal method of resolving a dispute. Such possible methods are the following.

a)   Trouble-shooting. The parties may agree to appoint an independent expert as soon as a problem arises. This trouble-shooter will consider the problem in the light of what the parties have agreed and decide on a course of action which is intended to solve the problem before it requires a more formal dispute resolution procedure. The parties can provide for the appointment of a trouble-shooter in the contract itself or agree to do so at the time the problem arises.

b)   Mediation. As an alternative, a mediator may be appointed. Such a person is usually an independent third party who provides a communication link between the parties and makes suggestions on how to resolve the dispute. The position of the consulting engineer in the FIDIC Rules is similar to that of a mediator, except that the impartiality of the engineer is considered to be doubtful by many observers.

c)   Conciliation. The conciliation procedure which the parties may choose
     is more formal than the other procedures, but is still a non-binding
     way of trying to resolve disputes. A conciliator is appointed who
     considers the arguments of both parties through written submissions,
     documentary evidence and, if he considers it necessary, an oral hearing.
     Once this examination of the problem is completed the conciliator will
     prepare proposed terms of an amicable settlement which he then gives
     to the parties for their consideration. As the conciliators' proposals
     are only suggestions and not orders, the parties are free to reject
     them.

     Many institutions which have arbitration facilities also provide
     conciliation procedures. For example, the ICC also has Rules of Optional
     Conciliation that provide for a conciliator to be appointed by the ICC.

d)   Concilio-arbitration. When the parties are deciding on the manner in
     which a dispute will be resolved, i.e., either at the time they are
     negotiating the commercial contract or when a dispute actually arises,
     they are also able to choose a concilio-arbitration procedure. Under
     this procedure, no party may make a request for arbitration until the
     parties have attempted to settle the matter amicably by a form of
     conciliation procedure the steps of which have been outlined in the
     agreement.

e)   Mini-trials. A system developed in the United States is the so-called
     mini-trial. The parties hire a private judge who conducts the trial.
     The parties agree that the time allowed for the hearing and the documents
     and other evidence submitted shall be limited.

f)   Summary arbitration. This is a form of arbitration which enables a
     dispute to be resolved as quickly as possible. There are no pleadings
     besides the statements of claim and defence and documents may only be
     submitted by the lawyers during their oral submissions to the arbitral
     tribunal and not before the hearing.

g)   Pre-Arbitration Dispute Resolution. This method of pre-arbitral dispute
     resolution appears to be gaining ground, albeit in various forms, in

engineering and construction, and arbitration bodies such as the
American Society of Civil Engineers, FIDIC, and ICC, although such a
procedure has been used by the European Space Agency for many years.
In general, this mechanism provides for rapid resolution of problems
prior to the commencement of an arbitration procedure. While it does
not necessarily preclude other forms of dispute resolution, e.g.,
arbitration or litigation, it appears to decrease the matters referred
on to further resolution by arbitration or litigation.

The purpose of all these non-binding procedures is to prevent the parties
from incurring the costs and spending the time involved with arbitration
on a dispute which is primarily technical in nature and thus may be more
satisfactorily settled in a less formal procedure, rather than through legal
proceedings. Most parties to engineering contracts are keen for its terms
to be carried out, such that a conciliation procedure or some form of
pre-arbitration dispute resolution may provide for a fast, economical and
non-adversarial resolution of a problem. Of course, it will only succeed
if all parties are interested that the matter be settled. If the ADR fails,
the parties are free to take the matter to arbitration or, in the absence
of an arbitration agreement, to revert to the courts.

REASONS TO ARBITRATE

When deciding whether a dispute should be resolved by litigation or a form
of arbitration all considerations should be borne in mind and there are,
of course, benefits and disadvantages to all. The main benefits and
disadvantages of arbitration compared to litigation will now be discussed.

1.   Benefits of Arbitration

a)   Confidentiality.

Unlike many court systems in which a dispute between parties becomes public
from the time a claim is filed, the entire arbitral process is usually
private. Most arbitration rules specifically state this privacy is
guaranteed. For example, Rule 2 of the Internal Rules of the International
Court of Arbitration of the ICC states:

"The work of the Court of Arbitration is of a confidential character
which must be respected by everyone who participates in that work in
whatever capacity."

Details of an arbitration may only be publicised if both parties consent.

b) Speed.

Particularly in the construction and engineering industry, speed in dispute
resolution is of the essence. An unresolved dispute may delay or even halt
a project which may be in varying stages of completion and everyone is losing
money with each working day.

As stated previously, the parties to an arbitration, in particular an ad
hoc arbitration, have a large degree of control over the procedural rules
governing an arbitration. Accordingly, the parties can generally control
how quickly a dispute will be resolved. It is thus possible for a dispute
to be resolved within days or weeks if the parties decide on the appropriate
steps - such as having no hearing and just written pleadings.

Alternatively, in a different situation the parties may decide it would be
quicker to have only an oral hearing with no written pleadings. The parties
can also agree that the arbitrator make only an oral, and not a written,
award or that it is not necessary for the arbitrator to give a fully reasoned
award, i.e., an award where the arbitrator states in full the reasons for
the conclusions he makes in his award.

The arbitrator may hear witnesses and accept documents or submissions, both
oral and written, in more than one language, if the parties agree. This means
there is no time lost by having the documents translated or witnesses heard
through interpreters.

An arbitrator, when accepting his nomination, undertakes to give priority
to the arbitration and such time and attention as is necessary and required
by the parties to have the dispute resolved as quickly and efficiently as
possible. This duty cannot be expected or demanded from the court when hearing
a matter. A claim filed at a court will be one of many which will have to

be dealt with by the court and will be affected by the other duties of the officers and judges of the court.

Courts in common law countries, such as England, the US, Canada and Australia, usually have one lengthy hearing rather than a series of shorter hearings. As a result, a trial date may be changed at the last minute because other hearings are taking longer than planned. Because of an already over-crowded hearing schedule the new hearing date is often months later. Consequently, the dispute will remained unresolved for some time.

A claim filed at a court will also be subject to set court procedures and time limits. It is only in extreme or special cases that a court will change these procedures and limits.

In court litigation, all documents not in the language of the court will have to be submitted with certified translations. Witnesses must be heard through court-certified interpreters. These are expensive and time-consuming procedures.

c) Expertise.

As the parties are able to choose the member or members of the arbitral tribunal, they are able to ensure that the persons who are to resolve the dispute are experts in the relevant field. Choosing an expert should signify that the person hearing the dispute already has a good understanding of its commercial and technical background. Thus, it would not be necessary to explain technical terms and concepts or to appoint an expert to assist the arbitrator. Obtaining proof by an expert witness is one of the main causes of delay in state court proceedings.

In court proceedings there can be specialisation to a certain extent in that the court system is often divided into e.g. commercial, family and criminal courts. However, the specialisation for commercial matters requiring technical expertise is generally rather limited and the presiding judge for a dispute is chosen by at random, or by rota, or by the first letter of the name of the defendant, and not because of any particular knowledge of the matter in dispute.

As a consequence, an expert is often appointed to assist the court with what may be highly complicated and specialised details and market practice. The appointment of a specialised arbitrator who is paid directly by the parties will generally in the end be more cost-effective than using the court system in which the initial court filing fees may be nominal or relatively small.

d) Neutrality.

With arbitration, it is possible for the parties to appoint arbitrators who are not from the same country as either of the parties and to choose both procedural and material laws which will not particularly benefit or disadvantage either of the parties. The parties are thus given every opportunity to obtain an arbitral tribunal which is above national laws and is unbiased with regard to the nationality of the parties and the law to be applied.

e) Finality.

Arbitration awards are, with few limited exceptions, final and thus there is little opportunity - or risk - of further proceedings once the award has been rendered. Enforcement can therefore be immediate; whereas a court judgment may involve two or three appeals to higher courts, which is of course very expensive and results in delay in final settlement and enforcement.

f) Enforceability.

There is now almost worldwide enforcement of commercial arbitration awards. The 1958 New York Convention on the Recognition and Enforcement of Foreign Arbitral Awards has been adhered to by most countries, including Austria. Other conventions such as the 1961 European Convention on International Commercial Arbitration and the 1979 Inter-American Convention on Extraterritorial Validity of Foreign Judgments and Arbitral Awards extend and reinforce the number of countries which enforce arbitral awards. Problems with enforcing an award will therefore only arise if an award is made, or enforcement sought, in a country which is not party to one of the conventions.

In comparison, the enforceability of a court judgment is much more limited

because it depends on bilateral and multi-lateral conventions and is affected
by local law procedural requirements. While an arbitral award made in Austria
is enforceable worldwide, a judgment given by an Austrian Court is enforceable
in about 15 countries but, unfortunately, not in important export countries
such as the USA and Japan.

2.   Disadvantages of Arbitration

a)   Consistent clauses.

Each contract must have its own agreement to arbitrate and an arbitration
will relate only to the contract which contains the relevant arbitration
clause. Other contracts, which may be otherwise commercially connected with
the contract, will and must be disregarded.

Third parties who are involved in the underlying commercial arrangement,
but are not party to the particular agreement containing the arbitration
clause may not become a party to the arbitration. This occurs frequently
with sub-contractors. Limited exceptions apply in the unusual case in which
all parties agree to join otherwise independent arbitrations.

The consequences of this are:

(i)   For each agreement, no matter how connected, there must be a separate
       arbitration clause;

(ii) Unless the arbitration clauses for each connected agreement are
       identical, it is possible that each one will be decided according
       to different material and procedural laws.

Courts, on the other hand, can assume jurisdiction over connected agreements,
can decide to have all disputes heard together and may apply consistent
principles, laws and rules.

b)   Reluctance to use.

There are certain areas of international commercial practice where there
is a considerable reluctance to have disputes resolved by arbitration. Banks,

in particular, are traditionally reluctant to become involved in arbitration proceedings. For example, international bank guarantees will rarely contain arbitration clauses. This is despite the fact that they are issued by banks upon the instructions of customers in relation to commercial contracts which will usually have arbitration clauses.

This results in the commercial agreement being resolved by arbitration and the bank guarantee issued in accordance with the terms of the contract being resolved by a court which may be in a different jurisdiction. Some new precedents of bank guarantees for the construction industry recommend the inclusion of arbitration clauses in the guarantee forms.

c) Agreement to arbitrate necessary.

Whereas to a considerable extent courts may assume jurisdiction, arbitration is considered a voluntary process and will therefore only be used when the parties have agreed to use it. Without this agreement it is impossible for any arbitral tribunal to accept jurisdiction.

However, although the agreement must be in writing, it is not necessary that it be in the commercial contract or that it be in one document. The agreement may be added to the contract later as an addendum or by subsequent agreement and by way of exchange of letters, faxes, telexes or EDI messages.

d) Local applicable procedural laws.

It is always possible that the arbitration will be subject to mandatory local procedural rules which can hinder the arbitral process in a way unforeseen by the parties. Care must be taken to investigate the local procedures in the event of their application.

REASONS FOR THE DIFFERENT TYPES

As discussed above, the parties, when choosing arbitration, will also have to decide on whether they will use institutional or ad hoc arbitration. Particular circumstances may make this decision easy; for example, parties to a contract for the supply of plants or utility facilities to state agencies are likely to opt for arbitration in accordance with the arbitration rules of the International Centre for the Settlement of Investment Disputes. Otherwise, the parties will have to decide whether it will be institutional or ad hoc, and then which rules to apply.

1.   Institutional Arbitration

Institutional arbitration is conducted in accordance with the requirements of a supervisory or administrative body. This body regulates the arbitral procedure by providing arbitration rules and supervising the proper conduct of the arbitration, e.g., by ensuring time limits are complied with, and some (i.e. the ICC) even check the award before it is submitted to the parties.

This regulation by the body generally ensures a greater degree of clarity and reliability to the arbitration process and the award eventually rendered. It is probably most appropriate to inexperienced parties who require a certain amount of protection and guidance in the conduct of the arbitration.

2.   Ad hoc Arbitration

On the other hand, ad hoc arbitration may be more suitable to the experienced party. A greater degree of control by the parties is possible. Without the formalities of an institutional arbitration (e.g., in an ICC arbitration there must always be prepared a "terms of reference", which must be signed by all parties and arbitrators before being able to proceed with a determination of the dispute by hearing or otherwise), an ad hoc arbitration is generally faster and more flexible. Without the costs payable to the institutional body, the costs are often considerably less.

The UNCITRAL Arbitration Rules have greatly facilitated the use of ad hoc arbitration and their inclusion in an ad hoc arbitation clause is highly recommended.

THE ARBITRATION CLAUSE

The consequences of a poorly drafted arbitration clause cannot be under-
estimated. Some of them have been mentioned above. The following are some
suggestions on contents of an arbitration clause.

1.   Type

The type of arbitral process chosen will affect the drafting of the clause.
Most bodies which provide arbitration rules also have a standard clause that
provides for a dispute to be resolved in accordance with its arbitration
rules. When such a clause exists, it should be followed because it will ensure
that the arbitration will proceed under the rules. However, most standard
clauses only provide for the basic agreement to submit a dispute under the
contract to arbitration in accordance with the specified rules. Further
consideration must be given by the parties to other clauses which should
be included in the agreement to arbitrate.

2.   Number of Arbitrators

The parties should consider how many arbitrators they would want to have
hearing the matter. As this is a rather difficult question, it is often
sensible to ensure that the parties have the option to have one or more
(usually three) arbitrators. This allows for flexibility when a dispute
arises. For example, a peripheral or simple dispute will be resolved more
quickly and cheaply if only one arbitrator is appointed. If the agreement
allows for only three arbitrators, then such a minor dispute could become
a lengthy and expensive matter.

On the other hand, if the contract is of relatively minor importance or does
not involve considerable amounts of money, it may be more advisable to
specifically allow for only one arbitrator and, in the case of disagreement,
have the arbitrator appointed by an independent third party. If the option
to have more than one arbitrator does not exist, this will then prevent one
party from insisting on three arbitrators as a delaying tactic.

3. Choice of Law

Choice of the material law will give certainty to a contract and reduce the costs and time involved in having this point determined at the time a dispute arises. It is otherwise often a very complicated issue.

Parties to international contracts often choose the law of a jurisdiction which is neutral to both countries. Switzerland and Austria are chosen because of their political neutrality and stability and often for their position in central Europe. However, before an unknown law is chosen, the wisest course is to obtain the advice of a lawyer in the given country as a precaution.

4. Language of the Arbitration

If the parties do not speak the same language, the contract and documentation are in another language and the arbitration to be heard in a country with yet a different language (a scenario which is quite possible for international construction and engineering contracts), there is the potential for a dispute to arise just to determine in which language the arbitration should be heard.

For the sake of clarity and when there is some doubt about which would be the most suitable language to be used, this point should be resolved initially and included in the arbitration agreement. Another benefit of this is that it should ensure that the arbitral tribunal will be capable of understanding the chosen language. It is then less likely interpreters will be required for the arbitrators and this should mean that no party will be more advantaged (or disadvantaged) over the other.

5. Place of Arbitration

The choice of the place of arbitration, which is the place where the hearings take place and where the arbitrators render the award, can be an important one.

Particularly with ad hoc arbitration, this may determine the procedural rules which will apply to the arbitral proceedings. In an ad hoc arbitration, if there is no agreement between the parties as to the procedural rules, then local law will determine the applicable procedural rules. Even if insitutional arbitration is chosen, in some cases, local law of the place of arbitration may provide overriding procedural rules. These can be of great significance.

As I have mentioned before, there are usually few problems with enforceability of an award because of the number and extent of the arbitration conventions. However, it should be remembered that any award, including an international one, is deemed to be an award rendered at the place where the arbitration is heard. It is therefore important that the selection of the place of arbitration will allow subsequent enforcement. Provided that the place chosen is in a country which is signatory to a suitable convention on the enforcement of arbitral awards, such as the New York Convention, there should be no problems.

14 A ROLE MODEL FOR THE VALIDATION OF MATHEMATICAL MODELS AND SOFTWARE INTEGRITY WITH RESPECT TO PRODUCT LIABILITY

ANUP PURI
GEC Marconi Defence, Warren Lane, Stanmore, England, UK

SUMMARY

Many organisations use validation procedures to a certain degree as a means of controlling accuracy and standards in their products and operations to meet Product Liability laws.

The paper presents a role model for a validation system designed to fulfil such requirements. Validation aspects of the following topics are covered; management of analysis operation; acquisition, development and verification of software; qualification and documentation of analysis methods; education and training of personnel.

1. INTRODUCTION

This paper states the principles that can be applied in the validation of computer models (mathematical) and software used in the design of products. The objective is to ensure that such models are validated to a degree appropriate to their intended use and are applied in a consistent and controlled manner for analysis affecting its integrity.

These requirements apply to any analysis type, which contributes to establishing the integrity of Grade A design, or to any analysis which is to be supplied as a deliverable in its own right where it may in future be so used.

2. FITNESS FOR PURPOSE

It is not a requirement that every computational analysis has to be performed to the highest degree of accuracy. The

Structural Failure: Technical, Legal and Insurance Aspects. Edited by H.P. Rossmanith. Published 1995 by E & FN Spon, 2-6 Boundary Row, London SE1 8HN, UK. ISBN: 0 419 20710 4.

key to effective computational analysis is to match the degree of control to the purpose of the analysis. To this end we can define three categories of importance of analyses as either VITAL, IMPORTANT or ADVISORY. These relate to the consequences of failure of the product and the role which the analysis fulfils in the demonstration of its integrity.

3. **OVERVIEW**

The total computational analysis system is modelled in terms of three serial activities:

- Acquisition, development and verification of analysis and associated software.

- Development and qualification of analysis methods; product design development.

- The management functions which implement and co-ordinate the activities; education, training and experience of personnel.

4. **ACQUISITION OF SOFTWARE**

Software may be developed in-house, developed by a third party to the Design Authority's requirements (subcontracted) or purchased as commercial off-the-shelf packages.

The Design Authority should ensure that software used in the computational analysis conforms with the application analysis requirements, particularly in respect to the analysis type. In acquiring software the Design Authority should define, where possible, the technical requirements and the tests or conditions which demonstrate satisfaction of those requirements.

The validation issues relating to purchased software are:-

- Evaluation of the suitability of the software to the analysis requirements of the product and the inherent limitations of the software.

- Verification of the software against its functional specification.

- The software suppliers Software Quality Control system.

- The software technical support service.

4.1 **Evaluation of Software**

Software evaluation is part of the product design organisation's quality system in software acquisition. The two sources of preliminary evaluation are the software theory and validation documents. Examination of the theoretical basis and numerical algorithms together with a range of bench mark problems may be sufficient to demonstrate the applicability and limitations of analysis software. Evaluation provides the technical management with the required appreciation of the software limitations.

4.2 Verification of Software

Software verification is primarily a software suppliers function and should be part of the suppliers Software Quality Control system, however, it is the responsibility of the user to ensure that the verification has been performed. The verification tests must include benchmarks tests (4.1). Basically these demonstrate that the software has satisfied fundamental tests for soundness and convergence - example, element tests for invariance, rigid body modes etc. The tests are examined for complete coverage of the theoretical basis, facilities and numerical algorithms.

4.3 Suppliers Software Quality Control System

The most important aspects of the software Suppliers Quality Control System (SQC) relevant to the product user are:

- Control of the software development process.

- Verification of the software.

- The procedures for identifying, controlling and correcting software errors.

Control of the software development insures against deterioration of the software under continual enhancement. The software quality control system must define the supplier's procedures for detecting, reporting, controlling and rectifying errors.

4.4 Software Support

The requirements for support of software used in integrity demonstration analyses include:

- Error and correction notification

- Documentation

- Training specific to the software

The software supplier must provide training in the use of the software. This should not merely cover the mechanics of input, but should include such topics as the solution methods used, limitations of the theory, element formulations and algorithms, diagnostic and error messages and their meaning, outline of the programme structure and operation.

4.5   Internal Software

All internally developed software used in qualified analysis procedures is subject to the same requirements as external software.   Use of uncontrolled software is identified as a significant risk.

4.6   Software Identification and Traceability

The release, issue or version of software used to produce each analysis should be identified and recorded.   Changes resulting in a new release issue or version should be recorded so as to produce traceability of software evolvement and of software used for different analyses.

5.   ANALYSIS PROCEDURE DEVELOPMENT

The development of qualified analysis procedures starts with a systematic review of methods in common and established use.   Documentation and qualification of those methods forms the basis of the analysis procedures library.

5.1   Documentation of Analysis Procedures

The documentation of an analysis procedure includes the following items:-

- The output data for which has been qualified and the order of its accuracy.   This is limited solely to the data correlated in the qualification analysis.   The order of accuracy is based on the degree of that correlation.

- The scope and limitations of its applicability, defined in sufficient detail to ensure that the procedure is not used outside that scope.   Typically these include the identifying characteristics of the physical model, analysis type, limitations of the behavioural modelling and theory assumptions.

- Reference to documented analysis used in qualification of the procedure.

- The maximum grade of importance of analysis in which the procedure may be used.   This is based on the degree and number of reference validation analysis. The grade of importance defaults to 'advisory' if the procedure is not qualified.

- The input data required for satisfactory execution of the procedure.

- The software and facilities to be used, e.g. mathematically modelling, solution techniques etc.

- The analyst controlled procedures, the level of detail must be sufficient to enable an analysis, of the relevant level of competence, to execute the analysis in a satisfactory manner.

- The QA checks to be exercised within the procedure, pre-analysis checks, analysis execution checks, post-analysis checks.

- Identification of the 'procedure owner' responsible for action on errors, omissions and queries.

5.2 Qualification of Analysis Procedures

Analysis procedure qualification is performed by analyses of realistic engineering models for which results, appropriate to the purpose of the analysis, can be confirmed by some independent means. Confirmatory results are typically obtained from physical tests supporting integrity demonstration of previous products, service experience, failure investigations, alternative analyses or third party assessments. Many test results can be found in open literature. The qualification tests are executed to the defined procedure, preferably by an independent analysis of the appropriate level of competence, to ensure that the procedure documentation is adequate. The documentation of a qualification analysis is sufficient to enable it to be repeated without reference to any other documentation so that, if the procedure is amended, the qualification can be repeated with a minimum of delay. Qualification analysis reports are subject to formal issue and change control. The software execution decks used in the qualification analysis, are added to the test library for software acceptance.

In assembling the qualification in respect to the grade of importance, both the degree of correlation and the reliability of the confirmatory values are taken into account. The more qualification tests and correlations available, then the greater the confidence in the procedure and the higher the grade of importance of analysis in which the procedure may be used. Qualification of analysis procedures provides project management with an appreciation of the inherent assumptions and limitation of the analysis method.

6. PRODUCT ANALYSIS

The product analysis is controlled by the project analysis manager through the project analysis plan. The individual analysis tasks are allocated through analysis specifications which use the analysis procedures qualified in the methods development activity. After checking of the results each analysis is documented, approved by the analysis project manager and used to update or amend the plan.

Details of the individual items are given in the following sections.

6.1   Project Analysis Plan

The project analysis is a dynamic document which starts as a broad outline and becomes more detailed as the design evolves. It is updated and amended as analysis results are obtained and may involve tasks in the methods development activity. At completion of the project the analysis plan becomes a record which correlates the individual analysis reports. The analysis plan is a controlled document subjected to periodic reissue and change control procedures.

Particular QA features of the plan are:-

-   The identification of decisions or reviews, based on structural features and analysis results, which may result in updating the plan and redirection of analysis activities.

-   Quantative assessments, estimates, design reviews and correlations between analyses to be used in checking the results from individual analysis.

-   The scope and grade of importance of each analysis.

Based on the scope of the analysis and its category of importance, each analysis task is allocated to an analysis team (Supervisor, analyst and software consultant) who collectively, fulfil the requirements of experience and expertise.

6.2   Analysis Specifications

The analysis team first prepares an analysis specification which is agreed by the project manager before the analysis proceeds. The purpose of the specification is to ensure that the analysis is sufficient to fulfil its purpose, that the appropriate input data is available and that the results are relevant to the project needs. The specification includes:-

-   The purpose of the analysis and output data required.

-   The sources of authentic data for input.

-   The qualified analysis procedure to be used.

- The input and results checking procedures to be invoked.

Where a qualified procedure is not available the procedure itself is outlined in the specification and the results used only for advisory purposes.

6.3 Results Checking

At completion of the analysis the results are checked and assessed. This involves:-

- Confirming that all the analysis procedure quality assurance checks have been executed satisfactorily.

- Comparison of the results with the estimates and correlations identified in the analysis plan. Quantative estimates are obtained from traditional simplified analysis.

- Assessment of the results based on knowledge of the physical problem. For analysis of high grade of importance, an independent assessment is performed by a qualified individual who is not a member of the analysis team.

7. Analysis Documentation

The controlled documentation of the analysis consists of an analysis report, the analysis record and project computer data files.

The analysis report provides only the information relevant to product design and integrity. It includes:-

- The purpose of the analysis.

- An outline of the representation of the physical problem by the analysis model.

- Summary, discussion and accuracy assessment of the principle results.

- Relevance of the results to the engineering problem and design recommendations,

  and

- References to enable further details to be obtained from the analysis records.

Summaries of analysis reports, including the scope of analysis, grade of importance and identification of analyst, supervisor and software consultant are stored on a database. This is used for subsequent task allocation decisions.

The <u>analysis record</u> includes:-

- The analysis specification.

- The key input data, in terms of the physical model and mathematical model representation, and the sources of that data.

- Selected output relevant to the purpose of the analysis.

- A summary of QA checks.

- Location of the input in the project computer files and the version/release number of the software used.

The analysis record, together with the stored input files, should be sufficient to enable the analysis to be repeated or updated reliably with a minimum of effort.

The <u>project computer files</u> contain the program input, and where appropriate, output relevant to the analysis. Where a number of design iterations are involved, only that germane to the definitive version is stored. The data to be retained is defined in the job closedown procedures and is stored in secured files.

8.   PERSONNEL REQUIREMENTS

The requirements are quantified in terms of:-

- Formal academic or professional qualifications.

- Product analysis experience.

- Mathematical modelling and problem solving relevant to the scope of the analysis, and

- Relevant software application experience.

The required degree of training and experience varies with the category of importance of the analysis and must be relevant to the scope of the analysis. Formal training in computational methods and software application contributes to personnel accreditation. Job experience, accumulated in one category, contributes to qualification for tasks of a higher category. Thus an analyst may perform low category tasks, under supervision, until sufficient experience is gained to perform such tasks unsupervised. He may then move to a higher category under supervision, and so on.

In cases where the required software application expertise is not available within the product design organisation, it may be necessary to contract suitable personnel from the software suppliers organisation. Similarly if the organisation is subcontracting analysis of an unfamiliar product, the necessary product expertise may be provided by the contractor.

8.1   Training should be provided by a combination of formal tuition and on job training. It is effective to use two types of course, one in general computational methods and others in the use of particular specialist software packages. This leads to a better appreciation of the analysis technology and maintains flexibility in the use of different codes.

8.2   **Personnel Records**

The personnel information should be extracted from the following sources:-

- Professional qualifications and years of engineering experience from the company personnel records.

- Training course attendance from the training records.

- The category of importance, scope of analysis, identification of the supervisor, analyst and program consultant, from the analysis report summaries.

This information should be co-ordinated in a data base and used in a variety of ways:-

- In task allocation. Given the category of importance, scope of analysis and a list of available personnel, all possible teams which fulfil the requirements are returned. This provides a management aid in task allocation and job scheduling. Note that this does not prevent the project manager from rejecting any team which, based on other considerations, is deemed to be inadequate.

- In the provision of training courses. Given the training course type, returns a list of personnel that require the training in order to progress to higher category of importance tasks. Provides a means of scheduling courses.

- In on job training. Given personnel identification, returns list of analysis scope, category of importance and roles, required to advance to higher category tasks. Provides a means of monitoring the advancement of individuals.

9.   **DISCUSSION**

The real quality of any computational analysis is determined by technical considerations. It is important to use experienced technical staff in determining computational procedures and using specialist Quality Assurance consultants in an advisory capacity.

The recommended first steps in implementing a computational validation system are to:-

- Constitute a technical body with responsibility and authority for the analysis validation system.

- Agree responsibilities, with the organisation, for each activity required in the system.

- Review current practices against the requirements.

ACKNOWLEDGEMENTS

The author acknowledges the valuable contributions made to this subject by members of MOD-JTRC Mathematical Modelling Working group and NAFEM-QA group, both of which the author is a prominent working member.

REFERENCES

1. I. Babuska.   "Reliability of Computational Processes in Structural Mechanics" - 6th World Congress on FEM, October 1990, Banff Canada.

2. J.J. Kacprzynski.   "Reliability Studies of Finite Element Methods in North America". 1987.   AGARD Report No. 784.

3. A. Puri.   "Capability, Accreditation of Personnel Competence in Organizations carrying out Structural Analysis using Finite Element Methods" - 3rd Int. Conf. on Structural Failure Product Liability and Technical Insurance.   Vol.2, Nos. 1/2 P280, 1990.

4. Nafems. "Quality Systems Supplement to ISO 9001 relating to Finite Element Analysis in the Design and Validation of Engineering Products - 1990, Dept. of Trade and Industry, NEL - Glasgow.

5. ISO 9001 Part 1.   "Quality Systems, Part 1 Specification for Design/Development, Production, Installation and Servicing".   1987.   British Standards Institution publication.

6. A. Puri.   "Role of Finite Element Analysis in Structural Qualification" - 6th World Congress on FEM Oct., 1990 Banff - Canada.

7. W.C. Rheinboldt & I. Babuska on "The Reliability and Optimality of Finite Element Methods".   Computer & Structures Vol. 10, 1979, 87-94.

8. R.R. Jackman and P.S. White "Review of Benchmark Problems for Nonlinear Material Behaviour". Oct 1987 - NEL Glasgow.

9. M.A. Chrisfield, G.W. Hunt and P.G Duxbury. "Benchmark Problems for Geometric Nonlinearity" Oct 1987 - DTI-NEL Glasgow.

10. J. Barlow. "Quality Management in Finite Element Analysis" Draft paper for AGARD, 1990.

11. K.J. Bathe, "Some issues for reliable Finite Element Analysis" Reliability of Methods of Eng. Analysis. pp 139-159, 1986.

12. A. Puri. "Finite Element Methods - A Black Box Approach to Structural Clearance". Proceedings of 4th Int. Conf. on Structural Failure & Product Liability p260-67 July, 1992.

13. J.W. Zing. "Survey of Results of Finite Element Quality Control". Finite Element News, June, 1989.

14. R.I. Wilson & I.J. Lloyd "Software Quality - Some Legal Issues" Proc. IFIP WG 5.4 Working Conf. on Approving Software Products (ASP-90). Sept. 1990 - Elsevier.

15. D. Sepahy & A. Puri - Draft Paper on Mathematical Modelling Analysis Procedure Development - MMWP - MOD 1993.

15 THE BRAZILIAN EXPERIENCE IN INTEGRITY EVALUATION

HENNER ALBERTO GOMIDE
Universidade Federal de Uberlandia, Uberlandia
SERGIO NELSON MANNHEIMER
Attorney at Law, Rio de Janeiro, and
TITO LUIZ DE SILVEIRA
Faculdade de Engenharia Souza Marques, Rio de Janeiro, Brazil

The three authors are members of the CTSL/ Comite em Technologia, Seguros e Legislação (Technology, Insurance and Legislation Committee) of the Associação Brasileira de Ciências Mecanicas (Brazilian Mechanical Sciences Association).

SUMMARY

The first organized Brazilian nucleus specifically dedicated to inspection for safety of in-service pressure vessel and piping was established in the petroleum industry. This happened in the mid-fifties. The experience achieved has been used as a model for other industries around the country. This paper presents a brief history of the development of this activity and its evolution towards inspection for integrity evaluation. General aspects of Brazilian law related to labor safety and civil liability are also commented. The authors express their views on the future needs in these two fields considering the Brazilian industrial panorama.

1 - CONCEPTS

Integrity evaluation is understood to be the activities of planning, results examination and interpretation when applied to equipment or systems to characterize the evolution of the accumulated damage in service, with the aim of guaranteeing the low cost preservation of safety and operational reliability long term.

Integrity evaluation represents no more than a label for the diagnosis and consequent maintenance recommendation within a modern concept of inspection for equipment in operation. The definition presented herein is general. It is applicable to productive systems in different technological universes. Even though they may have some basic principles in common, the specific surrounding conditions justify different actions in each case. Efficient solutions for the computers of a major banking conglomerate are not the same as those that fit to an airline company, even though the general safety and reliability aims at a low cost are similar. This work discusses the integrity evaluation applied to boilermaking equipment at continuous operation installations.

Structural Failure: Technical, Legal and Insurance Aspects. Edited by H.P. Rossmanith. Published 1996 by E & FN Spon, 2–6 Boundary Row, London SE1 8HN, UK. ISBN: 0 419 20710 4.

The typical industrial environments are: petroleum refining, petrochemical, power and the other segments that process fluids under pressure.

The motivation for integrity evaluation is a result of the ethical need of attending to the operators' safety, of protecting the environment, of guaranteeing product availability to the users and to maximize shareholder profits. Such motivation has been growing worldwide in recent years due to pressure from society bent upon greater safety and quality. In Brazil, in the last three years, cases of forced stoppages at large installations occurred, waiting for court decision on issues related to their employees' safety. Technicians responsible for inspection and for safety are being civilly and criminally sued as a result of accidents with victims. Another stimulus for integrity evaluation is the commercial nature. The opening of the market and the economic reality have imposed new competitive demands among companies. The viability of continuous operation installations tends towards jeopardy if the boilermaking equipment gives cause for frequent or prolonged un-programmed shutdowns. This risk is marked at old installations. The working life of large installations in Brazil has already reached a considerable extension and, as a result, many of them need substantial investment in modernization, be it to correct a natural accumulation of damage in service, or to get around the process' obsolescence. In each case the type and size of the most convenient investment and the verification of its opportunity should consider the physical state of the installation. Integrity evaluation is a diagnosis stage that is indispensable for the economic planning of actions that reduce the risks and the associated losses.

2 - ANTECEDENT

Integrity evaluation in boilermaking started to be considered in North America and Europe for equipment that operated at high temperatures and that were subject to accumulation of creep damage. In this case the mechanical design considers a finite life and adopts a reference life whose extension depends on the specific application. For the petroleum industry 100,000 h (11.4 years) is a typical value. For power generation this value can reach 250,000 h (28.5 years). The mechanical design is conservative and the result is that at the end of the reference life most boilermaking equipment is to be found in good state of repair requiring, if at all, localized repairs. Determined codes established, already in the 70's, that the equipment designed to operate in creep conditions would be submitted periodically to integrity evaluation once the reference life had expired [1].

The first nucleus specifically responsible for carrying out inspection on equipment in Brazil was formed in the Presidente Bernardes Refinery/ PETROBRAS in 1958 in the wake of accidents that had occurred. These groups were immediately set up in the rest of the refineries and later implanted into the petrochemical companies that were incorporated in the the 70's, as well as in some large companies in other areas. At that time, the activities that today are understood to be integrity evaluation were limited to the follow-up of the accumulation of corrosion damage. The characteristics specific to Brazilian industry in the 60's forced the model to specialize in boilermaking and to have a strong performance in maintenance quality control. The refining industry park was going through a period of accelerated expansion and the majority of the SEIEQs - Setores de Inspeção de Equipamentos (Equipment Inspection Sectors) absorbed responsibilities in the manufacturing inspection area at the suppliers, in assembly, in receiving materials and equipment, besides functions peculiar to the mechanical design and maintenance engineering, specially when referring to welding procedures and materials specifications.

The period between the 60's and 70's underwent various important advances. Highlighted, in 1962, was the formation of the Equipment Inspection Commission by the IBP -Instituto Brasileiro de Petroleo (the Brazilian Institute of Petroleum). In the 60's, this group published close to ten Inspection Guides with essentially a didactic aim. They made up the first documents in public circulation produced in Brazil on the subject. Another important contribution by the same group is the series of Equipment Inspection Workshops. The 19th event of this series is forecasted for May 1994. Throughout the years, this Commission has been the main forum in Brazil for the spreading of inspection technology for equipment in operation. Two other important advances resulted from PETROBRAS initiative. At the beginning of the 70's, a group responsible for the production of technical specifications for in-house use was installed, fulfilling the needs for national documents applicable to the petroleum industry. A variety of these standards covered wide gaps in the boilermaking inspection field and started to be adopted by other companies. In 1979, PETROBRAS set up an internal area for the qualification of personnel in distinct inspection modalities, also involving the performance of non-destructive tests. The initial aim was to take care of their own personnel and the contracted companies' employees for the rendering of inspection services. The results were excellent and the PETROBRAS qualification became the pre-requisite for innumerable companies.That mechanism made qualified labor available in the country and accessible for integrity evaluation projects.

3 - SETTING UP AND MATURING

The first cases of modern integrity evaluation date back to the second half of the 70', when the petrochemical companies started to evaluate the residual life of the catalytic reformer furnace tubes. In the first two cases, samples of the material were sent for analysis by manufacturers based abroad, that had access to the technology developed by the Battelle Institute [2]. Both results were frustrating owing to the excess of conservatism in the evaluations. Such a fact stimulated one of those companies to establish contact with universities. The reply was slow, as is common when dealing with the generation of new knowledge, since the Battelle experience was not available on that occasion. The result was the first Masters thesis produced at the Rio de Janeiro Federal University [3]. That is how the cooperation between companies and universities began, which has been defining the technological development in integrity evaluation and life extension in Brazil.

Integrity evaluations initially remained restricted reformer and pyrolisis furnaces. The first results outside this specific environment to the knowledge of the author were obtained between 1986 and 1987 in furnace coils fabricated from Cr-Mo steels at the Refinaria Presidente Bernardes / PETROBRAS and Petroquimica União. In 1987, the Manguinhos Petroleum Refinery carried out an integrity evaluation project for the life extension of a tower which at that time had been in operation for 240,000 h. Table 1 brings together some bibliographical references on the jobs carried out by the different companies. They illustrate the planning strategies,the results and the costs incurred. Integrity evaluation activities and the life extension have increased sharply at the beginning of the 90's, with the petroleum and petrochemical industries still leading the field. This is a common factor in Latin America and Brazil is no exception to the rule. This does not coincide with European and North American experience, where activity is centered around the power generation. Inasmuch that the philosophy may be the same in these two fields, the damage accumulation mechanisms are distinctly different. In that way, transference of experience established for steam to the petroleum and petrochemical environment demands adaptations and the utilization of specific technology. In Brazil the development of this technology has been achieved by cooperative effort between the companies that use the equipment, universities and renderers of services, some of them in some way linked to the universities.

TABLE 1 - INTEGRITY EVALUATION EXAMPLES		
Installation	**Company (Location)**	**Year**
Piratininga Power plant: four boilers with total power of 472 MW [4].	ELETROPAULO (São Paulo)	1988 - 1989
120 t/h Boiler, 525 deg.C, 120 kg/cm2 part of the utilities system of a refinery [5].	PETROBRAS / RPBC (Cubatão / SP)	1990
Liquefied Petroleum Gas Spheres [6].	PETROBRAS / RPBC (Cubatão / SP)	1989 - 1990
Main steam piping: pressure 128 kg/cm2, temperature 538 deg.C, material ASTM A 335 Gr.P22, diameters between 254 and 406 mm,wall thicknesses 33 to 62 mm, length 375 mm [7].	COPENE (Camaçari / Ba)	1993

It is equally important to register the role that the Technical Associations have played in contributing to the dissemination of experiences among companies. Besides the aforementioned IBP's Equipment Inspection Commission's initiatives, it is worth highlighting the actions of the ABCM - Associação Brasileira de Ciencias Mecanicas (the Brazilian Mechanical Sciences Association) that promoted various events as of 1983. In 1988 the ABCM set up an Integrity Sub-Committee. This group now became responsible for the organization of events in integrity evaluation and life extension to be held every two years and also responsible for the production of Recommended Practice documents. These documents do not have legal force, since the ABCM has no normative attributes, but has been given ample support by the equipment inspection community.

4 - ABCM RECOMMENDED PRACTICES

Until now three practices have been approved after debate involving the interested community. The main aims and the principle characteristics are summarized below. Availability of documents of this nature is fundamental for the activity's dissemination in the industrial environment and other documents should added to the group.

4.1 - RECOMMENDED PRACTICE ABC/CTVP-PR:001-92: General Recommendation for Equipment Inspection and Installations that Operate at High Temperatures.

This document is aimed at orienting the inspection of equipment potentially exposed to the action of damage accumulation mechanisms at high temperature so as to attend to one or more of the following objectives:
a) Evaluate the integrity, assuring a defined operation period;
b) Estimate the remaining life;
c) Recommend the moment, the extensiveness and the procedures for maintenance;
d) Supply information for revamp projects.

In its different parts, this paper offers: a basic terminology; an indication of the damage accumulation mechanisms; recommendation on when an evaluation should be started; a methodology according to which jobs are divided into stages: a preliminary evaluation; field tests and complementary analysis; and, finally, the attributes of the reports produced at each stage.

4.2 - RECOMMENDED PRACTICE ABC/CTVP-PR002-92: Field Metallography.

Its aim is to supply subsidy for the surface preparation; attack, observation and metallographic documentation performed in non-destructive character on the equipment's accessible surfaces in the location where these are to be found. The field metallographic procedures are used for:

a) Identifying causes of a known defect or a fault already   occurred;
b) Orientating and controlling repair procedures;
c) Collection of data for integrity evaluation.

The document covers the selection of areas for observation; the environmental conditions for the field work; surface preparation; metallographic attack; observation and micrographic documentation; operational cares; qualifications of procedures and operators; and the conclusive report's attributes.

4.3 - RECOMMENDED PRACTICE ABC/CTVP-PR004/92: Evaluation of the Propensity to Accumulate Creep Damage Due to Utilization Factor.

This practice refers to the equipment submitted to high temperatures with one of the following aims:

a) Verify if the necessity to go ahead with a detailed analysis or not exists for the characterization of accumulated creep damage.

b) Give hierarchy to the parts of an equipment or to the equipment within a system, so as to attribute priorities when programming an inspection.

Usage factor is defined as the sum of ratios between the fraction of the operating time and the minimum creep rupture time for all operating conditions, when applied to a certain part of the equipment and that these be defined from its history.

5 - ASPECTS OF BRAZILIAN LEGISLATION

Applied technology brings with it, intrinsic risks. Examples where the existence of an inherent risk is not able to condemn an activity or project are endless. Driving a vehicle in normal city traffic exposes the motorist, his passengers and the people around to a certain risk. The vehicle's manufacturer finds himself involved, according to his contribution to the worsenning of said risk. Brazilian legislation takes on a preventive, reparative and punitive character. In this way, traffic laws exist to reduce risks. Other laws state that a victim of being run over should be aided and awarded compensation. The driver should be punished if proved that he was the culprit. The punishment is worsened if it was premeditated.

In industry, prevention is considered in the Labor Laws and the legislation related to the environment. Boilermaking equipment are considered among the legal documents that regulate safety are boilers, pressure vessels and furnaces. The oldest document as regards this subject dates back to 1970. It is Decree No.20 passed by the Labor Ministry's Nacional Department for Hygiene and Safety that regulated the installation and inspection of boilers. This document contained many technical imperfections and its application to large scale industrial environments was problematic. In 1977, Law 6514 was published altering Chapter V of Title II of the Labor Laws related to Work Safety and Health. In its Section XII that law speaks of boilers, furnaces and pressure vessels. The following year the 28 Regulatory Standards (NR) foreseen in said law were approved by the Labor Ministry, of which two refer to equipment safety: NR-13 : Pressure Vessels and NR-14: Furnaces. Despite its title, NR-13 was applied exclusively to boilers, being almost a copy of Decree 20. In 1983, N-13 was revised with its applications being extended to: "other pressure vessels, like: compressors, compressed air tanks, compressed air cylinders, compressed air tanks (in general) and others, such as autoclaves that are as dangerous as boilers". The text's poor technical quality illustrates well the difficulties of the legal document's application. Revisions were introduced in 1984 and 1985 and a further one is presently being done. Since its origin, NR-14 has a laconic text that doesn't refer to inspection in operation, being limited to a few rules applicable to installation. Some of the most significant NR instructions are shown below:

NR - 1 : General Arrangement

* Establishes that: "the NR, relating to work safety and medicine are to be followed obligatorily by private, public and state companies and by direct or indirect public administration organs, as well as legislative and judiciary departments that have employees ruled by the Labor Laws".

* It is up to the Regional Labor Department (DRT): "to carry out the activities related to work safety and health and the supervision of the execution of the legal precepts and rules covering work safety and health". The DRT is empowered to solicit penalties, embargoes and interdiction due to the non-observation of the Regulatory Standards from the Civil Courts.

* NR-1 also establishes that it is up to the employer to obey and to see that the legal and regulatory rules are obeyed, on risk of application of the penalties foreseen in the pertinent legislation. It is up to the worker to follow these rules; unjustifiable refusal to do so constitutes a punishable act.

NR - 3 : Embargo and Interdiction.

* Establishes conditions for the embargo on jobs or the interdiction of facilities, due to a technical statement and opinion that shows serious and imminent risk.

* The interdiction and embargo can be solicited by the DRT's Work Safety and Medicine Sector or by a Trade Union.

NR - 13: Boiler and Pressure Vessel.

* Establishes that safety inspections, both internal and external, on boilers should be performed when new, after operational interruptions and periodically at least once a year, extendable for 6 months, by means of written justification sent to the DRT and based on a technical statement issued by the responsible engineer.

* This inspection must have a mechanical engineer, registered at the DRT's Regional Department, as responsible and it must follow the official technical standards in force in this country. The official standards are those issued by ABNT - Associação Brasileira de Normas Tecnicas (The Brazilian Association of Technical Standards). The only presently existing standard on the inspection of equipment in operation is NB-55: Safety Inspection of Stationary Waterpipe and Flame-pipe Boilers which has recently recently revised.

* Copies of inspection reports on boilers should be sent to the DRT and the Trade Union that represents the workers, that could demand inspection action from the DRT if any abnormality be found.

* NR-13 determines that " tanks for gases under pressure and compressed air tanks should be submitted to hydrostatic testing every five years or in the case of damage or repair. This test can be substituted for other tests by means of written justification sent to the DRT and based on a technical statement issued by the responsible engineer.

NR - 28 : Inspection Supervision and Penalties.

* Establishes, for each of the NRs, the items whose non-observation results in penalties. For each item the periodicity for correcting the situation are specified with the corresponding fines. The fines can be daily.

Damages and punishment arising from accidents with victims are regulated by the Civil Code and by the Penal Code respectively. It is necessary to distinguish the difference between guilt and malice with intent. The malice in this case is the directed intention to a determined result, in this case damage that could have been caused. Blame either exists or it doesn't. It exists or doesn't. The blame is a result of negligence, imprudence or incompetence. With negligence the subject proceeds in a rash manner omitting himself from an act that could avoid a damaging result. With imprudence he acts in a precipitated way, without his foreseeing the result of his conduct. Incompetence is characterizes the procedure of he who does not have the necessary qualifications for the performance of the job. As regards incompetence of a worker, the blame can very often be attributed to the employer, that when hiring the worker, or designating him to a specific task, could have given cause to the damage. The employer is charged with "culpa in elegendo". Defining the responsibility for the payment of damages can be complex. In this way the contracting company can be held responsible for damage caused by the contracted company if the contractor does not fully exercise the foreseen supervisory inspection and this omission be linked to the cause of the damage. It is called "culpa in vigilando".

The payment of damages according to Civil Law is subordinate to the principle of responsibility for the blame: he who causes the accident should opay for it once the blame has been proven. Circumstances exist which allow that the payment of damages be independent of the proof of the blame. One of these is worker activity, where damages are paid without it being necessary to prove the employer's guilt. It is sufficient for the injured worker to prove the connection between the work activity and the injury to have rights to basic compensation, this is called objective responsibility. In cases where objective responsibility, should the employer's blame be established, the injured worker accumulates basic compensation with a portion to be defined by common law. The question of compensation and punishment for possible blame, allows for an unlimited number of possibilities, depending on the interests involved.

As an example, the case where an operator of a production unit that, disobeying the internal norms and in an imprudent way, causes an accident of which the victim is a work colleague. Legislation presumes employer blame for the offending act of its employees or agents. In relation to the offending act, the victim can sue the company, or the imprudent operator, but the evident tendency is for suit against the company that has financial means for the additional compensation. At a later date the company could look for reimbursement from the imprudent operator by means of another suit. In parallel actions, the Public Prosecution could press suit against the imprudent operator and the Social Security Service could sue retroactively and regressively the company for reimbursement of the compensation paid by it to the victim.

As a second example, the production unit where top management has determined a reduction in the scope of activities recommended by the engineer responsible for the inspection. Let us suppose that the equipment stays in operation with undetected defects and, due to this, failed causing victims. The Civil Law suit possibilities are repeated for the compensation of the victims. To be exempt from criminal action moved by the Public Prosecution, the engineer responsible for the inspection must have at hand a document showing that the inspection's omission came from an order given by higher authority. Should he not be able to do so the blame is delineated as negligence or incompetence. The technical decisions' documentation inside the companies is frequently unclear as to responsibilities.

6 - COMMENTS

6.1 -A growing number of companies have been acquiring the instruments to manage the accumulated damage in service in their equipment. Revamps are increasingly more frequent that, besides correcting accumulated damage, introduce alterations in the process capable of achieving higher efficiency levels to those of the original design. The lack of resources inherent in an emerging economy, like the Brazilian one, increases the value to the extended life of the existing units. Integrity evaluation and life extension of pressure vessels and piping are activities that should continue to expand in quantity and quality throughout the 90's. As a reaction to this demand, companies that render specialized services and the active research and development institutions are multiplying.

6.2 - The accumulated damages that represent effective commitment for the integrity, in general is concentrated in small parts of the equipment, which is highlighted further when considering a unit as a part of the whole plant. As a result, the planning phase, in which one starts the integrity evaluation, should be selective when defining locations and the inspection methods. Independently of the strategy adopted by the companies that are owners of the process installations being more vertical or more distributive, with the intensive contracting of services, the integrity evaluation is a multi-disciplinary activity and its efficiency depends on the close involvement of its technical staff including operational and process specialists.

6.3 - As the integrity evaluation goes deeper in a large installation, the detailed analysis of the equipment allows that several thousand points of interest be defined. The volume of information tends to make inviable an evaluation, already in the preliminary planning stage. A new development frontier is the creation of computerized systems for data collection, storage and analysis. Systems for this purpose exist interlinked to process records that allow for real time data collection, aiding the analysis of damage that accumulates during fluctuations. There are specialist systems, in development stage, that incorporate analysis procedures and respective judgement criteria arranged in a decisions tree. This would seem to be the future for accumulated damage management in large installations.

6.4 - One concept should be clear to all that intend to make use of the consolidated knowledge in the literature on damage accumulation mechanisms as a means to accelerate their own actions, and shortening the roads to efficiency. The greatest part of available experience, be it in literature or stored in the memory of the renderer of services, refers to the steam environment and should go through adaptations if the scope being discussed involves other systems. The origin of the experience should be clear when its application is intended to a determinate system. Even within steam generation, there exist differences in behavior as regards the damage accumulation between European and American equipment. They can be attributed to different design conceptions. Among equipment of the same origin, even among equipment from the same manufacturer, important differences can exist between the damage accumulation mechanisms as a result of unequal operational regimes. In any case, the good practice of inspection engineering, together with the practice of multi-disciplines, are the requirements for integrity evaluation success. The integrity evaluation program managers should create space, in terms of time and resources, for the good practice of inspection engineering from the planning stage.

6.5 - In Brazil, at the end of the 80's, many traditional inspection sectors in the petroleum and petro-chemical industries demonstrated a certain enormity due to the absorption of functions typical of quality control and others but that are not intrinsic to inspection. The corporate needs for inspection evolved. The facts to be considered are: labor for boilermaking services have reached a much higher qualification level than that of the 70's; abundant offer for qualified services in non-destructive testing; new quality guarantee concepts, where the control actions are shouldered by the groups in charge of production; little attention from the great majority of the inspection groups to integrity evaluation activity. Following the restructuring wave that has swept the country in recent years, many companies' organization charts have been altered reflecting on the inspection sectors. These changes do not follow a uniform pattern. Even in PETROBRAS, origin of the classical model, the solutions vary among refineries. In several companies inspection of equipment in operation ceased to exist as a sectors, the responsibilities being distributed one or more of the remaining sectors. In some companies, inspection activity was abolished, with all the services being handed over to third parties. In others the scope was reduced with emphasis on integrity evaluation, the sector's status being elevated due to the shortening of the report distance to top management. As these alterations are relatively recent and the process hasn't finished, it is not yet possible to establish a dominant trend for the future. Whichever the path to be trod by inspection groups might be, integrity evaluation is attracting growing attention of the process companies in Brazil.

6.6 - It is the authors' opinion that the inspection for integrity evaluation be within the strategic activity within a process company, feeding the relevant managerial decisions that involve multiple commercial interests. The management of accumulated damage by boilermaking equipment is an activity that requires continuity. In this way the authors believe that every process company should count on a highly qualified inspection structure of its own. It should be structured in such a way that it should at least be able to manage the acquirement of data and to interpret the results of the analyses performed. The intermediary inspection stages could be more or less delegated to the contracted companies, depending on the general policy of the process company towards out-sourcing.

6.7 - In Brazil it is the inspection groups that, by tradition, form an opinion on the state in which boilermaking equipment is to be found and the registering of the damage accumulation history via recommendations and inspection reports. When legal suit is pressed involving the examination of responsibilities in an accident, the tendency is for the direct involvement of the inspection group. Following the rules pertaining to work safety is a second source of interaction of the inspection groups with the legislation.

6.8 - Brazilian legislation directed towards the safety of pressure vessels and piping safety presents problems that hinder its implementation in large process units. One of the problems resides in the generality of the scope to which the texts with legal value adhere. In relation to boilers, NR-13 states that: "A boiler is all the equipment destined to produce steam under greater pressure than the atmosphere, using any outside source of power." It is evident that the adequate actions to guarantee the safety of a small steam generator for use in a laundry, for example, are not the same as those required by a power boiler. Despite the problems, it is the general consensus that the legislation results are positive. It is recognized that the legislation has contributed to reducing the number of accidents, that in the past were much higher. In recent years, large companies have tried to obey the law. The legal texts have been perfected and NR-13 is presently undergoing revision.

6.9 - The full application of the requirements set out in NR-13 and NB-55 do not guarantee the safety of a boiler or a pressure vessel. The legal documents are clear when attributing, to the responsible engineer, the discernment on the majority of the inspection actions and the evaluation to be carried out, he being responsible for any faults arising from faulty inspection.

6.10 - Integrity evaluation, as defined, is an inspection committed to obtaining safety and reliability with maximum economic efficiency. The legislation refers to inspection aimed uniquely to look after safety and has within it no worries about cost savings. An inspection that states the unsafe of a pressure vessel or boiler and shows the need for a long repair shutdown, fulfills the requirements of the law but is entirely inconvenient from the economic point of view.

6.11 - Brazilian legislation governing worker protection is old. For a long time it remained half forgotten, especially by the large process companies, although they are in general the most efficient in protecting the safety of their employees. Among the changes introduced into the legislation over the years came the involvement of the communities interested in the inspection supervision. These communities, in this case, are represented by the Trade Unions. Disregard to the legislation exposes the company to fines. A condition of serious and imminent risk allows the DRT, or a Trade Union, to solicit the courts for the company's interdiction. The understanding of what a serious and imminent risk would be is subjective and the court's action is relatively slow in face of the large continuous process units' time scale. Some concern exists that this question may evolve into a wider conflict between companies and Trade Unions. The question as a whole is particularly sensitive, at this moment, bearing in mind recent cases occurred at a Liquified Petroleum Gas terminal belonging to PETROBRAS and at the Petroquimica União complex.

6.12 - The Civil Code imposes on companies the financial damages, for accidents at work, independently of proof of blame. Compensations increase if fault is proven. Those are the risks faced by the companies. Most of the large continuous process companies have a well established tradition for looking after their employees' safety. Many of the installations however are intrinsically dangerous, since they deal with fluids under high pressure and temperatures, some of them with hazardous fluids.Accidents happen [8]. Recently, growing unrest is perceivable among those technicians responsible for the inspection activities owing to the palpable possibility of criminal suit in the case of an accident with victims. The unfolding of the incident that occurred in 1992 at Petroquimica União has stimulated this reaction.

7 - CONCLUSIONS

7.1 - It is noted that the preservation of production facilities in safe, reliable and efficient conditions attends society as a whole and not only the companies that hold the usage of those installations. The preservation principle is a vital demand on the emerging economies as is the case with Brazil. It should be said that common sense diagnosis that is being discussed in this paper, continues to represent a challenge. We need to develop and spread technology that leads the diagnosis to encompass equipment types and industrial environments, besides those that have, until now, been contemplated.

7.2 - Technology for pressure vessels and piping integrity evaluation is to be found consolidated in Brazil with several projects finished and a number of others underway. In relation to the future, the integrity evaluation path in Brazil admits the following unfolding:

a) Deepen the fundamental understandings on particular damage accumulation mechanisms and transform that know-how into analysis criteria and inspection procedures;

b) Develop computerized systems capable of managing integrity evaluation in wide ranging equipment groups and no longer in a small amount of equipment treated separately.

c) Extend integrity evaluation to other equipment other than boilermaking, for example: rotational, electric equipment and so on, in such a way as to achieve the widest reach for the evaluations within each productive unit.

d) Disseminate and publish the available knowledge to greater number of companies and productive units.

e) Modernize and maintain the existing official technical standards related to safety and the environment.

7.3 - Brazilian legislation states that the prevention of accidents at work involving boilers, pressure vessels and furnaces since 1970. Not adopting the regulations exposes a company to punishment and interdiction, should the court accept the solicitation from the DRT or the Trade Union based on serious and imminent risk. Cases that occurred recently incite worry with a possible conflict front. Legislation related to safety is considered imperfect but a consensus exists that it contributes to the reduction of accidents that were frequent in the past, mainly in small installations. A company is not obliged by law to practice the modern integrity evaluation, whose objectives are much wider than the operator's physical safety guaranteed by the Legislation.

BIBLIOGRAFIA

[1] - BS5500 Specification for Unifired Welded Pressure Vessels, British Standard Institution, London 1976;

[2] - Dillinger, L.; Roach, D.; Metallurgy and Metallography of Reformer Alloys; Battelle Memorial Institute, Columbus, Ohio, Estados Unidos, 1975;

[3] - Paiva, R. L.C.; Características Mecânicas e Estruturais de Tubos de Aço Inoxidável 35% Ni - 25% Cr. Utilizados em Fornos de Reforma Catalítica, Tese de Mestrado, COPPE - Universidade Federal do Rio de Janeiro, 104 p. dez. 1980;

[4] - Ortiz, J.N.G. et allii; Extensão de Vida Útil de Usinas Termoelétricas. Conceito e Aplicações; Anais do ICV 91 - Simpósio Nacional Sobre Integridade em Centrais de Vapor; ABCM ; Nova Friburgo, ago. 1991;

[5] - Brambilla, P.A.; Ramos, E.A.S.; Avaliação de Integridade de uma   Caldeira e sua Linha de Distribuição; Anais do 18° Seminário Brasileiro de Inspeção de Equipamentos; IBP; pp. 967-980, Rio de Janeiro, jun. 1991.

[6] - Bernardes, R. et allii; Avaliação da Integridade Física das Esferas de GLP da Refinaria Presidente Bernardes - PETROBRAS; Anais do 18° Seminário Brasileiro de Inspeção de Equipamentos; IBP; pp 981-992; Rio de Janeiro, jun. 1991.

[7] - Sales, A.A.; Avaliação de Integridade de Tubulação de Vapor de Alta Pressão e Alta Temperatura da UTE da COPENE; Anais do IEV 93 - Conferencia Internacional Sobre Avaliação de Integridade e Extensão de Vida de Equipamentos Indústriais; ABCM; pp 203-208; Pouso Alto, set. 1993.

[8] - Garrison, W.G.; Major Fires and Explosions Analyzed for 30 Year Period; Hydrocarbon Processing, pp. 115-120, set. 1988.

16 AGEING, SAFETY AND LEGAL ASPECTS OF CABLEWAYS IN AUSTRIA

T. VARGA and E. CORAZZA
Institute for Research and Testing of Materials (TVFA), Technical University
Vienna, Austria

Austria has many mountains, but still more holiday resorts. Tourists have to be transported in most cases up to the peaks. Cableways of different type are used generally. There are over 2700 in service. Their long term safety is therefore an important problem.

One may distinguish three groups of components for cableways: those, which are subjected to Periodic recurring inspections (non destructively), those which have to be type tested and accepted and finally those, which are not bound to any formal regulation, e. g. recurring inspection.

A typical representative component for the periodic recurring non destructive inspection by magnetic testing, using saturation (where the secondary magnetic field of the cable indicates cracks, notches or thinning of the wire) are both fixed suspension or traction cables are inspected, see Fig. 1.

The maximum length of an inspection interval is 4 years; with degradation of the cables shorter inspection intervals may be recommended by the laboratory who is in charge of the examination. This laboratory has to be at present authorised by government; as the latest by 1995 it has to have an accreditation by an appropriate body.

Subject to type tests are clamps, see Fig. 2, cabin or seat suspension parts (Figs. 3 and 4) and cabins or seats themselves (as in Fig. 4). The most important examinations concern fatigue: loads between the maximum service load and three times the maximum service load have to be applied 5 million times.

Structural Failure: Technical, Legal and Insurance Aspects. Edited by H.P. Rossmanith. Published 1996 by E & FN Spon, 2–6 Boundary Row, London SE1 8HN, UK. ISBN: 0 419 20710 4.

Fig 1: A two seat cable car hanging on the cable (af-ter it was derai-led)

Fig 2: Fatigue testing of a clamp, inclination variable between + 45 and - 45 degrees of angle.

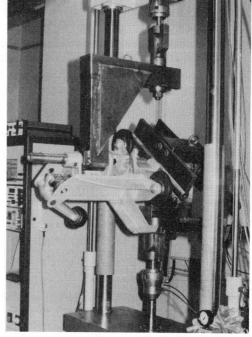

Fig 3: Fatigue testing of a connecting member, suspension of a cable car

Fig 4: Cable cabin, prepared for fatigue testing

If after completion of 5.10^6 cycles, cracks are found during visual examination, than the structure in question (like in Fig. 5) has not met the test criteria. There are cases, where changes in design, material selection and of fabrication care plus quality control are needed to improve the fatigue properties as they are needed in the part depicted in Fig. 6. This happens time by time both in suspension parts as well as in clamps, specially to those, which are not fixed to the cable permanently. In the latter case clamping-release-clamping cycles are applied.

One of the traditional questions is, how the spring stiffness changes, if one of (for instance disk shaped) springs is fractured. If all the different conditions as included in the requirements are fulfilled, the fitness for purpose is assured and the laboratories in charge may recommend acceptance by the authorities.

Finally, even on such components, which are not subject to a recurring inspection, ageing phenomena may be followed up by examinations which are agreed upon between the laboratory in charge and the company using the cableway. For instance for simple, permanently fixed clamps magnetic particle testing is conducted on 1/10 of the overall number of clamps applied on the cableway in question. A fatigue crack which could have been detected, and the consequent brittle fracture of the clamp, is demonstrated in Fig. 7.

If no cracks are found at all, the clamps are o.k. in respect to fatigue. If cracks are found, even tiny ones in dangerous positions, than a further examination of all clamps of the cableway is required.

Another ageing mechanism, corrosion has also to be dealt with: tube profiles are checked by ultrasonic testing (UT) concerning wall thickness; but also pitting corrosion is investigated. If the wall thickness drops locally below the necessary value, immediate action, i.e. complete revision of the tubular structures has to be carried out and non acceptable parts taken out of service, see also the tube section in Fig. 8.

The examples described above show already the close interrelation of technical examinations and the legal regulations: both together result in operational safety of the cableways.

Fig. 5: Connec-
ting members of a
cable car (bet-
ween roll battery
and cabin)

Fig 6: Fatigue cracking of
a forged part of a
clamp (after pene-
tration testing)

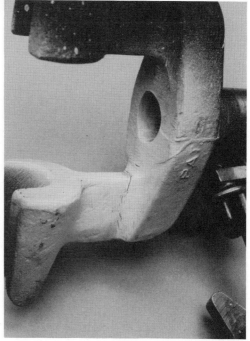

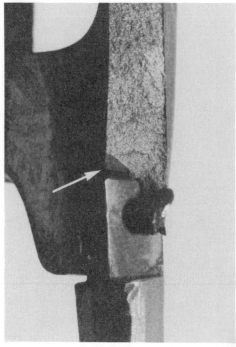

Fig 7: Fracture surface of a clamp, which originated from a fatigue precrack. The piece was broken in brittle manner, as the cable was torn out.

Fig 8: Thinning of the tube section due to corrosion on the inner surface.

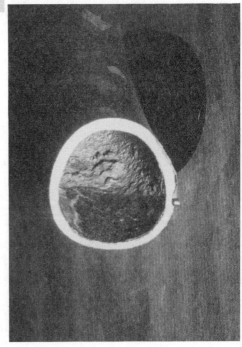

In very few cases, however, difficulties are observed. If regulations have been violated, no change of existing rules becomes necessary. In case no violation of the rules were observed, maybe the rules have to be made more stringent.

This was the case after the only accident in Austria, which occured in 1992. The cable has derailed due to the fracture of the side ring of a a cable roll (see Fig. 9) and subsequent torsional resonance vibration of a pillar. The first cableway rolls in a roll battery had to be changed and replaced by one of a better quality. Because this case is not finished before court, no further explanation is possible now.

After an accident, investigations become necessary. Soon after the accident, insurance will be asked for financial help. In an urgent case, where, as usual, different insurance companies are involved, one of the companies may furnish support without prejudice, i. e. the actual distribution of the responsabilities takes place later.

For an appropriate distribution of responsabilities to different parties, like for the equipment in question, responsabilities of the subcontractor(s) to the maker, the cableway traffic company and/or its employees, the forensic investigation is of utmost importance.

Therefore the investigating laboratory has on one hand to represent the know-how and have the necessary equipment in a sufficient measure at disposal, on the other hand it must be sufficiently familiar with the functioning of the equipment which led to the actual casualty. It must be also in picture concerning the valid regulations and rules. There will be only in special cases a single expert who can deal with such a problem in all relevant aspects. From metallography, fractography, non destructive testing, fracture mechanics, good design considerations, material properties, i.e. from the know how of a good, experimental laboratory, all aspects have to be covered to arrive up to realistic conclusions of the cause.

The insurance company expert has, on his part, to cover also all the different aspects mentioned. If the laboratory is trustworthy, than he may rely on its findings. However, it will be always preferable, if he is able to put forward an own critical appraisal of the laboratory results, especially which are connec-

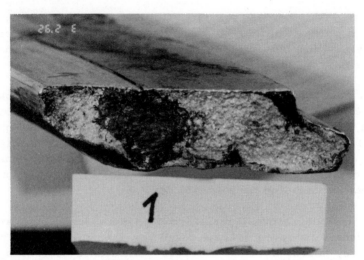

Fig 9: Broken aluminium ring on a roll. The fracture was acused by the large dark defect, which initiated the fracture due to the acting force. This fracture was the cause of the derailment of the cable.

ted with the duties of insurance companies and the legal regulations thereof.

The insurance covers not only the casualties, which have already happened, but the major part of its activity is dealing with damage, of each equipment which may happen, is also still at risk. Well managed insurance companies have developed and apply risk management philosophies. Now it depends on the basis, whether such philosophies have a positive impact on the insurance business of that company and the basic risk aspect may be treated in a reasonable way.

Are only phenomenological aspects taken into account, which in many cases have no relation to the selection of the equipment and its function, than the value of such a risk management is very limited. If, however, the design philosophy, the fabrication, testing and control, the function and possible malfunctions and their causes are taken into account, a major improvement in performance may follow. The synthesis of know how from design and mechanical functions, fabrication, control procedures and quality control, including ageing procedures and its consequences, will become decisive in risk management considerations.

Finally, a "closed loop" should be envisaged: risk management considerations should improve all the beforementioned safety factors before and during fabrication, also in maintenance and recurring inspections. By this a more exact input of data is guaranteed for the next risk analysis and in consequence, for risk management.

Literature:

T. Varga: "Grundsätzliche Betrachtungen zur Lebensdauer von
        Seilbahnen"; O.I.T.A.F. Seminar, 19.4.1991, Wien,
        p. 6.1-6.8
E. Corazza: "Zerstörungsfreie Prüfungen zur Gewährleistung der
        Betriebssicherheit von Seilbahnen"; O.I.T.A.F.-Seminar, 19.4.1991, Wien, p. 9.1-9.23
T. Varga, E. Corazza: "Aging in ropeways"; ISR 7/1993, p.27-32
T. Varga, E. Corazza: "On the ageing of cableways";
                        7. International Congress on Cableways,
                        2.-5.6.1993, Barcelona

17 MECHANICAL CHARACTERISATION OF PREFATIGUED 2219-T87 AND 6061-T6 ALUMINIUM ALLOYS UNDER HIGH VELOCITY TENSION

M. ITABASHI, K. KAWATA and S. KUSAKA
Department of Materials Science and Technology, Science University of Tokyo, Noda, Chiba, Japan

ABSTRACT

From economical aspect, commercial aircraft operators seem to want to extend the use−period of their aged aircrafts. The fuselage and wings of them include invisible defects due to fatigue phenomenon.

In the present article, combinations of fatigue and impact loading were examined to detect differences of mechanical properties for 2219−T87 and 6061−T6 aluminum alloys between virgin and damaged condition. Serious degradation of mechanical properties for pre−fatigued 2219−T87 aluminum alloy in dynamic tensile test was observed. According to these experimental results, the authors would like to sound the alarm for the extending use−period programs for the aged aircrafts uncritically under only economical consideration.

1. INTRODUCTION

Service conditions for structural materials have become severer and severer monotonously, since new technology has always been accompanied with the demand for

Structural Failure: Technical, Legal and Insurance Aspects. Edited by H.P. Rossmanith. Published 1996 by E & FN Spon, 2-6 Boundary Row, London SE1 8HN, UK. ISBN: 0 419 20710 4.

more sophisticated performance. Nowadays, once the machine or structure of high performance is broken by an accident, the damage tends to be of a large extent. The cause of such an accident is frequently seen to be the fatigue failure phenomenon.

So far, according to the concepts of damage tolerance or fail safe, S−N curve, i.e. stress level versus number of repeated loading at rupture relationship, was accepted as the most reliable guideline of the design for the machines or structures exposed to repeated loadings. The curve indeed gives only the lifetime of the material under expected service stress level to designers and engineers. However, for example, the fuselage of modern aircrafts was not designed under consideration of degradation or embrittlement of the materials during their own life−spans, before the appearance of visible defects on the components of them. Of course, such thin−walled structures including invisible defects are not said "safe".

Fatigue, as above mentioned, is one of the major causes of accidents. Another major cause is impact. An impact loading is not easy to analyze strictly, because two aspects should be considered simultaneously. Firstly, the dynamic force propagates along the structure components as a wave motion. Secondly, a material of the component behaves differently from its own quasi−static behavior, during the dynamic deformation. The former aspect is skipped in this article. The present argument focuses on the material behavior. The latter aspect needs experimental data to clarify the difference between the quasi−static behavior of the material and the dynamic one. Quasi−static tensile test for metallic materials had been standardized in each industrialized country. In case of dynamic tensile test, each investigator has originally−designed experimental apparatus[1]. The present authors operate a high velocity tensile loading machine of a horizontal slingshot type[2] adopting the one bar method[3]. This loading machine can afford the tensile velocity of 10 m/s typically.

So far, the fatigue and impact problems have been investigated independently. The authors wondered what would occur in the combination of two major problems. They have challenged this complicated problem.

2. A TYPICAL APPLICATION

At first, a typical application is introduced. Fig.1 is a map of Japan and neighboring countries. Japan is composed of four main islands Hokkaido, Honshu, Shikoku and Kyushu, and a lot of accompanying small islands. Dotted lines show a network of Shinkansen superexpress railway. A solid line shows an airway between Tokyo and Sapporo.

Between Tokyo and Sapporo 32 flights go and return in one day, as shown in Table 1. This scheduled airway corresponds to 90 min.—flight. 70 % of passeger seats were occupied through 1992[4]. 24 flights are operated with Boeing 747 every day. Boeing 747 is an aircraft type of intercontinental flight and huge capacity. This route is the most frequently scheduled airway in the world with large capacity aircrafts, such as Boeing 747, McDonnell Douglas DC−10 and Airbus A300. And don't forget this is a domestic route. Why does the Japanese people want such a large jet route? The reason is that we have no Shinkansen line direct to Hokkaido across the Tsugaru strait. If you choose a railway trip for the same destination from Tokyo station, you must spend more than 10 hours.

Usual daily operating condition of Boeing 747 is only 1 or 2 flight cycles, i.e. 1 or

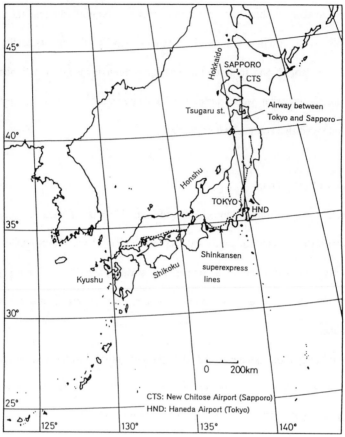

Fig.1   An airway between Tokyo and Sapporo and Shinkansen superexpress
railways

Table 1   Number of flights and aircraft types between Tokyo and Sapporo

TOKYO −−−−−− SAPPORO
(HANEDA, HND)   (NEW CHITOSE, CTS)
32 flights of 3 airlines (JAL, ANA and JAS) per day
90 min.   70 % seats occupied (1992)

Aircraft type	No. of flights	Capacity
Boeing 747SR or 747−400D	24	550
Airbus A300 B2/B4 or A300−600R	4	300
McDonnell Douglas DC−10−30	4	300
Total	32	10,100 per day

◎ If you choose a railway trip for the same destination from Tokyo station, you
must spend more than 10 hrs.

2 taking—offs and landings. On the other hand, in Japan, domestic 747s are operated in severer conditions, 3 or 4 flight cycles per day. In Table 2, it indicates the specifications of Boeing 747 family. There is the type of 747SR. SR means 'short range'. You cannot see this type in New York, Chicago, London, Paris, Amsterdam and Frankfurt am Main. This type was specially designed for Japanese domestic routes. The most special performance of 747SR is its flight cycles, almost twice of other types. Boeing reduced 20 % of fuel tank capacity, strengthened wings, fuselage and landing gears of 747SR, and then extended the flight cycles. Recently, one more special type named 747—400D was rolled out. D means 'domestic'. 400D was almost the same specifications, somewhat better.

Each flight cycle gives a certain loading pattern to the aircraft. One flight cycle is equivalent to a reduction of the remaining lifetime of it. This is very fatigue phenomenon. And, due to bad weather conditions, the aircraft experiences some dynamic loadings of unexpected level in a gust and hard landing. Then, the combination of the fatigue and impact is generated sometimes.

Nowadays, under the global economical recession, the commercial aircraft manufacturers and operators seem to be intending to extend the use—period of their aged aircrafts than the designed ones. Such use—period extending programs should be carried out under serious considerations of not only economical but also technical points.

Table 2   Designed lifetimes of Boeing 747 family

Aircraft type	Flight time, hrs	Flight cycle, times (No. of take—off and land)
Boeing 747SR	42,000	52,000
Boeing 747 of other types	62,000	24,600

Ⓒ There is one more domestic type for Japanese airline companies, named 747—400D (Domestic).

3. STRAIN RATE EFFECT

The present authors have contributed mainly to the field of high velocity deformation of solid materials, not the fatigue problem. Fig.2 is a stress−strain diagram of a typical steel(0.45%C). Generally speaking, for design, mechanical properties of structural materials have been obtained by standardized quasi−static tensile test. The terminology 'quasi−static' means almost stop moving. So, loading speed is very slow. However, in practice, traffic accidents on the road and collisions of aerospace structure against the ground or a meteoroid are very high velocity phenomenon. At that time, the difference of loading velocities exceeds 5 or 6 decades, frequently more. The structural material reveals their own strain rate effect. Mechanical engineers call it strain rate effect. In order to estimate such an effect, two types of loading machines must be used. One is a commercially−available universal material testing machine for the quasi−static test.

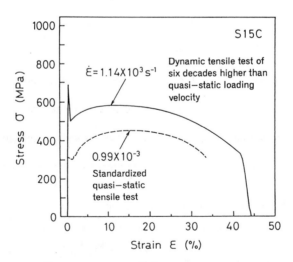

Fig.2   Strain rate effect revealed to tensile stress−strain curves for a carbon steel S15C (0.15%C) at quasi−static and dynamic loading velocities respectively

Another one is a specially—designed high velocity loading machine. One of the most famous high velocity loading techniques is the split Hopkinson pressure bar method. The original type was for axial compression[5]. Then, it was modified to axial tensile loading[6]. The authors utilize a more sophisticated technique, one bar method, based upon newly derived formulae, developed in 1979[3].

In Fig.3, these diagrams show mechanical behavior of an aluminum alloy and a steel (the same as Fig.2) at two different loading velocities. Coordinates indicate the tensile stress and abscissas the tensile strain, respectively. So far, in the top side of Fig.3, aluminum alloys were known as materials of a few strain rate effect. Broken line indicates the behavior under quasi—static tension, 0.5 mm/min. Solid line does under dynamic tension, 10 m/s, equivalent to 36 km/h. On the other hand, a typical steel shows obvious strain rate sensitivity, like the bottom side of Fig.3. The authors have accumulated such experimental results[7] for a lot of metallic[2,8], polymeric[9], inorganic[10], and composite materials[11,12,13]. But, they were all virgin materials. This time, they extended their interests to materials damaged by fatigue.

4. EXPERIMENTAL APPARATUS

In this series of experiments, two commercially—available testing machines and one originally—designed machine were used as shown in Table 3. First one is an axial fatigue testing machine. Second one is a universal material testing machine. A quasi—static tensile loading velocity was set 0.5 mm/min, which corresponded to the strain rate of 0.001 s^{-1}. The last one, shown in Figs.4,5, is a high velocity tensile loading machine of a horizontal slingshot type adopting the one bar method which has been

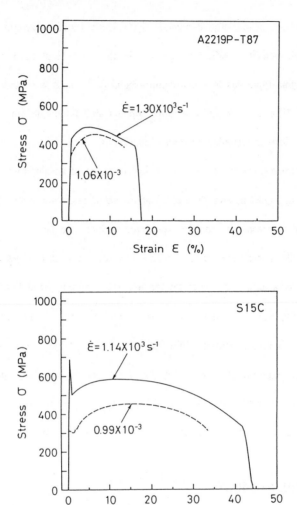

Fig.3   Tensile stress–strain diagrams for virgin 2219–T87 aluminum alloy and also virgin S15C carbon steel

Table 3   Experimental apparatus

① Axial fatigue testing machine

> Shimadzu Servo Pulser
> Type : EHF−FB1 1111
> Capacity : ± 9.8 kN

② Universal material testing machine

> Shimadzu Autograph
> Type : AG−10TA
> Capacity : ± 98 kN
> Loading velocity : 0.5 mm/min (in present study)

③ High velocity tensile loading machine of horizontal slingshot type with one bar method

> Originally designed
> Duration of measurement : 1160 μs
> Capacity : 20 kN
> Loading velocity : 10 m/s = 36 km/hr

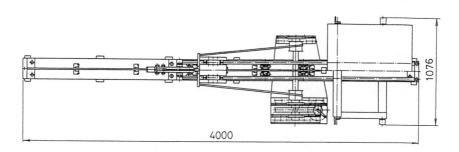

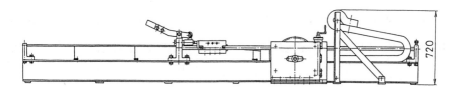

Fig.4   Assembly drawing of the high velocity tensile loading machine of a

horizontal slingshot type (unit:mm)

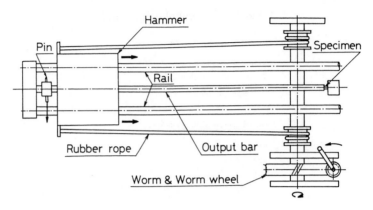

Fig.5   Schematic drawing of the high velocity tensile loading machine adopting

the one bar method

constructed by ourselves. A dynamic loading velocity was set 10 m/s, corresponding to

1000 s^{-1}, so 6 decades higher than the quasi−static one.

5. SPECIMEN

Chemical compositions of aluminum alloys are tabulated in Table 4. 2219−T87

alloy is Cu rich with good weldability and good stability at elevated temperature, used as

fuel tank of artificial satellites and is a candidate of pressure wall of United States' space

station. 6061−T6 alloy is Mg rich with good corrosion resistance and relatively good

crashworthiness, used as chemical rotors, automotive wheels and also a candidate of

bumper plate of space station against space debris and meteoroid impact.

Table 4   Chemical compositions

Material	Si	Fe	Cu	Mn	Mg	Cr	Zn	Zr	V	Ti	Others	Al
A2219P−T87	0.07	0.16	5.90	0.26	0.00	−	0.03	0.12	0.12	0.03	0.01	Bal.
A6061P−T6	0.7	0.25	0.30	0.12	1.1	0.19	0.01	−	−	0.01	0.01	Bal.

Specimen dimensions are relatively small as indicated in Fig.6, comparing with the standardized quasi–static tensile test specimen. The diameter is 3 mm and gage length 8 mm including round fillets of both sides.

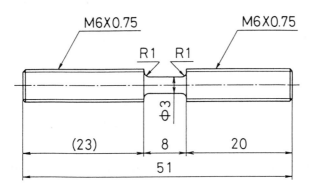

Fig.6   Specimen dimensions (unit:mm)

6. EXPERIMENTAL PROCEDURES

The present series of experiments was composed of three stages. At the first stage, the S–N curves of each tested Al alloy were obtained. A fatigue condition was indicated in Fig.7. Only tensile side, sinusoidal pulsating load was applied at 5 Hz cyclic frequency. According to the S–N curve, once the stress level is set at a certain value, the number of repeated loading at fracture is automatically fixed. This is a nature of the materials. The life of the material was also designed by setting the service stress level. In Fig.8, each solid line is the S–N curve for 2219–T87 and 6061–T6 aluminum alloys, respectively. Small circular symbols are experimental plots. The higher stress level is given, the smaller number of repeated loadings at fracture is obtained.

Pre–fatigue conditions were set as indicated large circular plots in Fig.8. The severest condition of the stress levels is 89 % of quasi–static tensile strength and 80 %

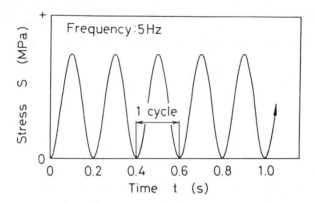

Fig.7   Fatigue condition for the S—N curve

of the number of repeated loading at fracture. There exists a certain statistical distribution of experimental plots around the S—N curve. During the preparation of the pre—fatigued specimens at the severest condition, the authors frequently failed to do, due to the unexpected fracture of the specimens. Anyway, at the second stage, the pre—fatigued specimens without visible defect were prepared. The pre—fatigue conditions were severer than those of actual design and service, in order to detect boldly something different.

At the last stage, the tensile tests of two velocities were carried out. Fig.9 is stress—strain curves of the virgin materials. The stress levels for pre—fatigue are also shown inside them. The higher level should be avoided in use at design. The lower level may be possible in very—tightly—designed structure. The purpose of the present article is how differently the pre—fatigued specimens behave between quasi—static and dynamic tension.

S : Stress level of pre-fatigue
$\overline{\sigma}_p$: Quasi-static tensile strength
n : Number of repeated loading
N : Number of repeated loading at fracture

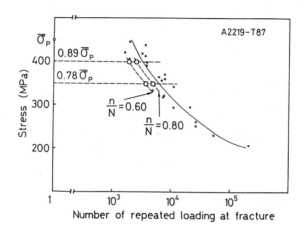

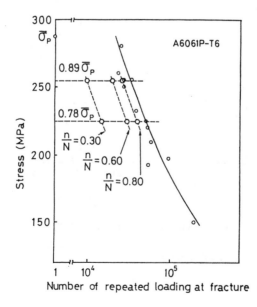

Fig.8 S—N curves and pre—fatigue conditions for 2219—T87 and 6061—T6 aluminum alloys

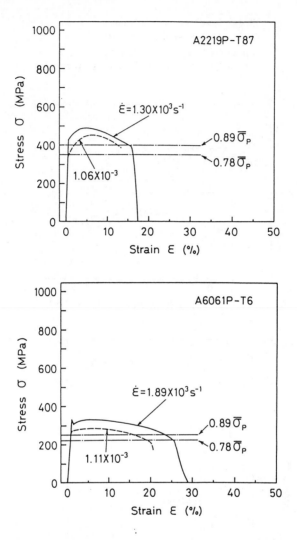

Fig.9   Stress–strain curves for virgin 2219–T87 and 6061–T6 aluminum alloys in quasi–static and dynamic tension

7. DEFINITIONS OF MECHANICAL CHARACTERISTIC VALUES

In the left side of Fig.10, mechanical engineers always pick up such mechanical characteristic values from a stress–strain curve as their routine work. The most important characteristic values are maximum stress and maximum strain. The maximum stress is termed as tensile strength or ultimate strength. The maximum strain is termed as total strain or breaking strain. It should be also focused on the enclosed area of the stress–strain curve and the abscissa as crashworthiness of the material. The area means the amount of energy which requires to break the material. So, the value is called as absorbed energy per unit volume.

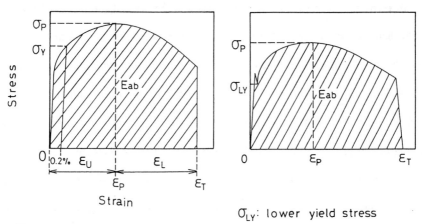

σ_{LY}: lower yield stress

σ_p: tensile strength
σ_y: proof stress
E_{ab}: absorbed energy
              per unit volume
ε_p: strain to σ_p
ε_T: total strain
ε_U: uniform strain ($=\varepsilon_p$)
ε_L: local strain ($=\varepsilon_T-\varepsilon_p$)

Fig.10   Definition of mechanical characteristic values

Proof stress is a limited stress which do not generate permanent strain in the material. This stress level is defined as the value at 0.2 % plastic strain. In dynamic tensile tests, such curves with yield—like behavior were frequently obtained as shown the right side of Fig.10. So, in case of such a curve, lower yield stress was taken instead of the proof stress.

8. EXPERIMENTAL RESULTS AND CONSIDERATION

Fig.11 indicates the effects of both pre—fatigue and strain rate in the tensile strength. The left side and right side correspond to 2219—T87 aluminum alloy, and to 6061—T6 respectively. In each figure, the abscissa is the pre—fatigue ratio, and the left half and right half correspond to the lower stress level of pre—fatigue and to the higher stress level of pre—fatigue, respectively. Circular symbols show the tensile strength in the dynamic tensile test, and square symbols the data in the quasi—static test. Square symbols are located on an almost horizontal line for both alloys at each stress level, and scattering of the plots is small. The fact means that, in the quasi—static tension, any pre—fatigue effect could not be detected. For 2219—T87 alloy, it degrades remarkably in the dynamic tension after the pre—fatigue. Mean values in each experimental condition are connected by solid lines in the dynamic tension and by broken lines in the quasi—static tension. A crossing of solid and broken lines can be seen for each stress level. These crossings are very, very dangerous and undesirable phenomena. On the other hand, 6061—T6 alloy is a relatively safe material. Circles are always situated above squares, even in the pre—fatigued conditions. Some dynamic tensile specimens show tremendously high tensile strengths. Maybe, these higher strength specimens are

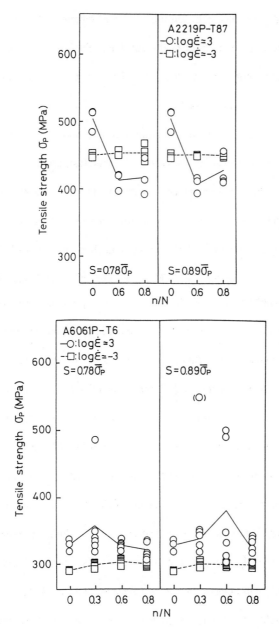

Fig.11   Pre−fatigue and strain rate effect in tensile strength for 2219−T87 and 6061−T6 aluminum alloys

influenced by slight misalignment during the pre–fatigue. Plots with parentheses correspond to rupture at round fillet of the specimen. Such part is stress concentrated and hardened during the pre–fatigue and some vacancies confirmed by optical microscope observation are introduced.

Fig.12 shows the pre–fatigue and strain rate effects in the proof stress or lower yield stress in the same mannar. In the case of dynamic test, proof stress or lower yield stress shows almost the same tendencies with the tensile strength in the previous paragraph. Comparing with the tensile strength, however, quasi–static symbols are relatively scattered. The crossing of the mean value lines for 2219–T87 alloy can be seen again and no crossing for 6061–T6 alloy. For 2219–T87 alloy just after the pre–fatigue, any visible damage on the surface of the specimens was not observed by the naked eye. Invisible damage is suspected to exist inside of the specimen.

Frankly speaking, the breaking strain or total strain is relatively scattered compared with the stress values. Fig.13 represents results for the total strain. For 2219–T87 alloy, quasi–static total strain is almost constant. Dynamic total strain maintains value comparable with quasi–static. For 6061–T6 alloy, the pre–fatigue effect in the quasi–static tests did not exist obviously. In case of the higher stress level, dynamic total strain indicates a critical tendency. That is, it once extremely falls and recovers to the quasi–static level. So, this fact suggests, during pre–fatigue process, microstructure is furiously changed at the higher stress level. And this fact was not detected without the high velocity tensile test technique. Formerlly, in the quasi–static test for pre–fatigued 2024–T3 alloy, degradation of the total strain was reported[14].

Fig.14 shows results of the absorbed energy per unit volume. This value corresponds to crashworthiness of the material in the dynamic test. 2219–T87 alloy,

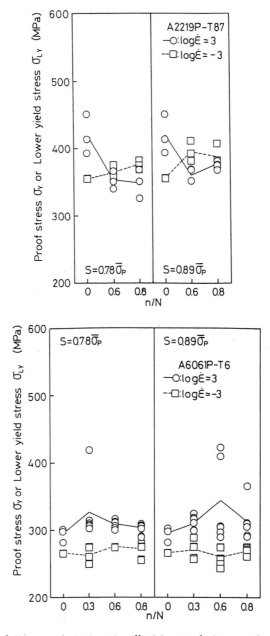

Fig.12 Pre−fatigue and strain rate effect in proof stress or lower yield strength for 2219−T87 and 6061−T6 aluminum alloys

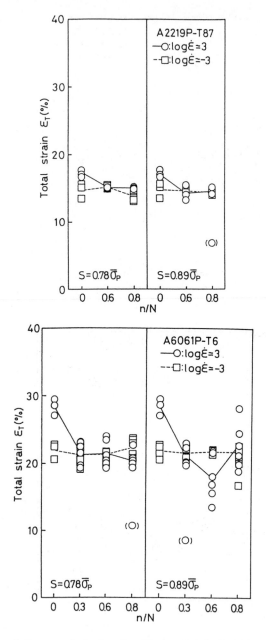

Fig.13   Pre−fatigue and strain rate effect in total strain for 2219−T87 and

6061−T6 aluminum alloys

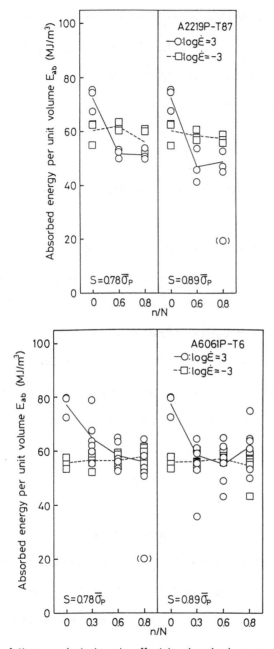

Fig.14   Pre−fatigue and strain rate effect in absorbed energy per unit volume for 2219−T87 and 6061−T6 aluminum alloys

after the pre−fatigue, is less crashworthy than in its virgin condition. For 6061−T6 alloy, at the lower stress level, the dynamic absorbed energy is comparable with the quasi−static one, except the specimen fractured in fillet. But, at the higher stress level, it shows rather complex tendency, down and up.

9. CONCLUSIONS

According to the experimental results including serious facts, it is concluded as follows:

(1)Residual mechanical properties for pre−fatigued aluminum alloys in the dynamic tension have been successfully obtained. Comparing with well−known properties in the quasi−static test, the dynamic experimental results include very serious differences.

(2)These problems will be investigated further with electron microscope technique.

(3)The present authors do not intend to blame aircraft manufacturers for their present design. But, it should be noticed that the state of the art is progressing, like this. They and airline companies should refer these results in the extending lifetime programs for their aged aircrafts.

REFERENCES

[1] Kawata, K. and Itabashi, M., "Analysis of Elastic Bar Method for Materials Characterization in High Velocity Tension," Elastic Wave Propagation, (Proceedings of the 2nd IUTAM−IUPAP Symposium on Elastic Wave Propagation, Galway, Ireland, edited by M.F. McCarthy and M.A. Hayes), Elsevier Science Publishers, (1989), 223−228.

[2] Kawata, K., Miyamoto, I., Itabashi, M. and Sekino, S., "On the Effects of Alloy Components in the High Velocity Tensile Properties," Impact Loading and Dynamic Behaviour of Materials, (Proceedings of IMPACT '87, Bremen, West Germany, edited by C.Y. Chiem, H.−D. Kunze and L.W. Meyer), DGM Informationsgesellschaft Verlag, Oberursel, 1, (1988), 349−356.

[3] Kawata, K., Hashimoto, S., Kurokawa, K. and Kanayama, N., "A New Testing Method for the Characterization of Materials in High Velocity Tension," Mechanical Properties at High Rates of Strain 1979, Inst. of Phys., Conf. Ser. No.47, (edited by J. Harding), Bristol and London, (1979), 71−80.

[4] Airline, Ikaros Publications, Tokyo, 170, (1993), 88, (in Japanese).

[5] Kolsky, H., "An Investigation of the Mechanical Properties of Materials at Very High Rates of Loading," Proc. Phys. Soc. Ser.B, 62, (1949), 676−700.

[6] Harding, J., Wood, E.O. and Campbell, J.D., "Tensile Testing of Materials at Impact Rates of Strain," J. Mech. Eng. Sci., 2, 2, (1960), 88−96.

[7] Kawata, K., Hashimoto, S., Sekino, S. and Takeda, N., "Macro− and Micro−Mechanics of High−Velocity Brittleness and High−Velocity Ductility of Solids," Macro− and Micro−Mechanics of High Velocity Deformation and Fracture, (edited by K. Kawata and J. Shioiri), Springer−Verlag, Berlin, Heidelberg, (1987), 1−25.

[8] Igata, N., Kawata, K., Itabashi, M., Yumoto, H., Sawada, K. and Kitahara, H., "Plastic Deformation of Iron and Steel at High Strain Rate," Advanced Materials for Future Industries: Needs and Seeds, (Proceedings of the 2nd Japan International SAMPE Symposium and Exhibition, Chiba, Japan, edited by I. Kimpara, K. Kageyama and Y. Kagawa), International Convention Management, Tokyo, (1991), 1121−1129.

[9] Kawata, K., Chung, H.−L. and Itabashi, M., "Mechanical Characterization and High Velocity Ductility of HTPB Propellant Binder," Proceedings of the 6th Japan−U.S. Conference on Composite Materials, Orland, USA, Technomic Publishing, Lancaster, (1993), 771−781.

[10] Kawata, K., Hashimoto, S., Miyamoto, I. and Hirayama, T., "On the Mechanical Behaviour of Inorganic and Organic Glassy Solids in High Strain Rate Tension," Composites '86, (Proceedings of the 3rd Japan−U.S. Conference on Composite Materials, edited by K. Kawata, S. Umekawa and A. Kobayashi), Japan Society for Composite Materials, Tokyo, (1986), 69−75.

[11] Ger, G.−S., Kawata, K. and Itabashi, M., "The Mechanical Properties in High Velocity Tension of CFRP, KFRP and CF/KF Hybrid Composites," Achievement in Composites in Japan and the United States, (Proceedings of the 5th Japan−U.S. Conference on Composite Materials, Tokyo, Japan, edited by A. Kobayashi), Kokon Shoin, Tokyo, (1990), 95−102.

[12] Kawata, K., Itabashi, M. and Fujitsuka, S., "Characterization of Highly Anisotropically Reinforced Solids in High Velocity Tension," Mechanical Identification of Composites, (Proceedings of European Mechanics Colloquium 269, 'Experimental Identification of the Mechanical Characteristics of Composite Materials and Structure,' Saint−Etienne, France, edited by A. Vautrin and H. Sol), Elsevier Applied Science, (1991), 231−237.

[13] Chung, H.−L., Kawata, K. and Itabashi, M., "Tensile Strain Rate Effect in Mechanical Properties of Dummy HTPB Propellants," J. Appl. Polym. Sci., 50, (1993), 57−66.

[14] Kawata, K., Hashimoto, S. and Hondo, A., "Mechanical Degradation of Aeroplane Materials by Their Fatigue and Its Detection (1st Report) −Decreasing of the Breaking Strain of 2024−T3 Al Alloy by Its Fatigue−, Report of Institute of Space and Aeronautical Science, University of Tokyo, 6, 3(B), (1970), 716−728, (in Japanese).

18 FATIGUE FAILURES AND IDENTIFICATION OF THEIR CAUSES

MATTHEW BILY
Slovak Academy of Sciences, Bratislava, Slovak Republic

Abstract
The main causes of fatigue failures are described, and the complexity of the influencing factors is discussed. Implications for designers and users of structural components are outlined, with examples from many fields of activity.

1 Introduction

Men during their evolution have suffered with many epidemic diseases. Although their battle has been more or less successful, there is certainly one, let me call it disease, which according to its widespread and economic consequences is still truly pandemic. It is the *fracture of various artificial and natural structures* such as ships, airplanes, vehicles, pipelines, pressure vessels, shoes, glass windows, teeth, bones, limbs, dams and thousands of other components.

It has been estimated, for example, that the losses in the American economy due to fracture amounted to 4% of the gross national product which makes about 1.9×10^{11} US Dollars. A similar situation is faced in the European Community with approximately the same order of the gross product. In less developed countries the situation is probably even worse as the losses may exceed the percentage of 4%[1]. These figures are really frightening especially when taken together with the social, psychological and moral effects of fatalities, injuries, mental health changes and the changed way of life which usually accompanies accidents resulting from a fracture.

There are various kinds of service fractures usually characterized as *brittle, ductile, creep* or *fatigue*. Although their frequency of appearance may vary depending on the type of equipment, there is no doubt that the prevailing reason for 70–90% of all service fractures is the fatigue damage evoked by repeated loading.

Why is it so or, in other words, why is it practically so difficult to prevent such numerous occurrences of fatigue fractures?

First of all, material and structural fatigue is an *extremely complex phenomenon* governed by 20–30 known and also some unknown factors acting

Structural Failure: Technical, Legal and Insurance Aspects. Edited by H.P. Rossmanith. Published 1996 by E & FN Spon, 2–6 Boundary Row, London SE1 8HN, UK. ISBN: 0 419 20710 4.

synergetically, as shown schematically in Fig. 1. Because it is practically impossible to formulate and realize a multifactorial experiment which could evaluate and quantify the partial influence of each factor, our designing rules are mostly based on the assessment of a single factor effect on the fatigue life. In some cases this may not be adequate or it may even be fatal.

Further, it is not easy to define the exact *use conditions* of a structure to be designed or to estimate the corresponding operating loads and environment. Considering as an example such a well understood item as a car, one should take into account road unevenness, speed, engaged gear, payload, road slope, driver's individual performance, maintenance effect, environmental temperature, corrosive environment and many other randomly appearing events. Should even all these factors be defined and estimated, it is more than difficult to specify the percentage of their mutual appearance in the typical use unit. A simple example of this procedure is documented in Fig. 2 on a lorry delivering goods and running empty. Here only four operating parameters, speed, load size, road and performance, have been considered with their corresponding two or three levels of severity, supposing that the environmental parameters are constant. It is obvious that the use unit is represented by 17 vectors with various percentages of their appearance.

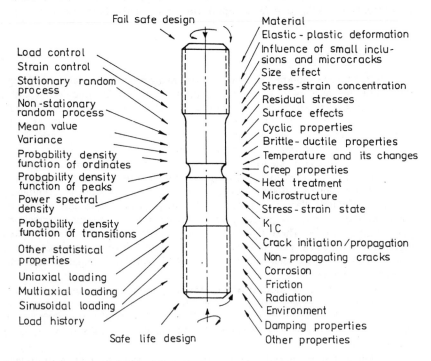

Fig 1: A review of possible factors playing a role in fatigue processes. Left hand side, operating load factors: right hand side, material and technological factors.

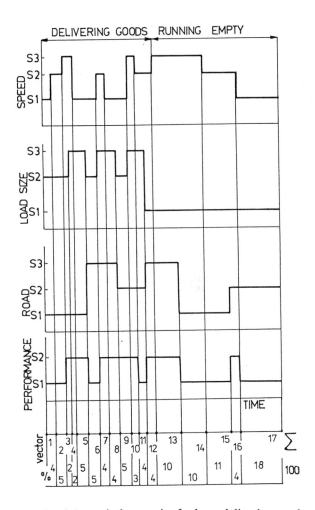

Fig 2: An example of the typical use unit of a lorry delivering goods and running
empty, composed of 17 vectors.

Numerous examples of collapsed bridges, crashed airplanes, road accidents, offshore catastrophes, power station fatalities, etc bring evident proof that the fatigue phenomenon under real use conditions still belongs to "terra incognita" and this is why fatigue fractures are not unusual. Such disgraceful events cannot, naturally, be tolerated and so their causes are to be disclosed to provide information for the designer, lawyer and/or insurer and at the same time to mark the victim who might cover or at least share the losses.

In doing so it is unavoidable to consider all possible or even impossible factors which could influence the whole process of fatigue damage accumulation. A review of them is in Fig. 3. They can be divided into a few groups.

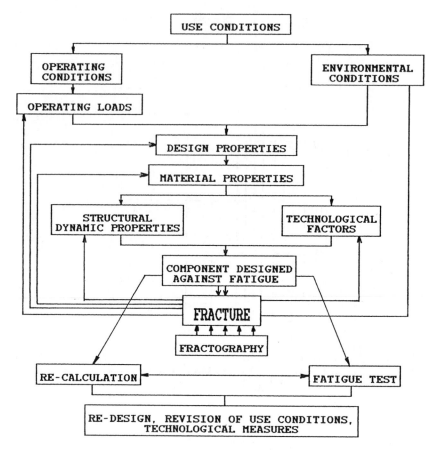

Fig 3: Groups of factors to be considered when investigating the causes of service fracture.

2   Use conditions

As argued above, the use conditions represent the primary input for the reliability (fatigue life) estimation, but they may be very complex and so it is hardly possible to consider all of them when creating a product. Thus it is wise to inquire of every possible person involved about extraordinary situations, unusual operating conditions (functional modes, input signals, loads, manipulation by operator, various supplies, etc) as well as environmental conditions (climatic, shocks, earthquake, lightning, biological agents, water splashing, atmospheric corrosion, etc) and/or their unexpected severities. Such a thorough questionnaire may give a clue to what happened and what operating loads might have led to the unexpected damage accumulation.

3   Design properties

Fatigue is very sensitive to any shape disturbance such as a hole, groove, changed cross-section, transition radius etc which causes local rise of the nominal stress σ_{nom} up to its peak value σ_{peak} exceeding the local fatigue limit σ_c (Fig. 4) and leads to initiation of the fatigue crack. Thus the fractured component should be very carefully examined from this point of view and the potential stress concentrations should be revealed. It follows from our experience that most service fatigue failures are due to this effect.

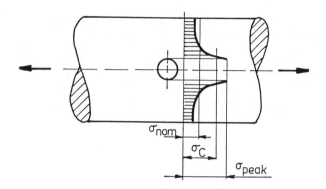

Fig 4: Nominal stress σ_{nom}, peak stress σ_{peak} and the fatigue limit σ_c in a component with a stress concentration.

4   Material and technological factors

Fatigue properties are uniquely determined by the cyclic material properties. Nevertheless, not those properties given in standards and reference books but the properties of the specific material used which are heavily influenced by strain ageing, heat treatment, plastic prestrain, loading mode (stress vs strain control), level of loading, etc. Thus *never trust the material supplier* if the application is expected to be critical and always try to perform your own fatigue tests or ask someone reliable to carry them out.

As fatigue is known to be a *surface phenomenon* (fatigue cracks practically always initiate in the surface layer) one should also be careful when deciding on the final surface treatment. A rule of thumb suggests that the better the material (e.g. high strength steel), the smoother the surface (e.g. ground or polished) required, otherwise the expensive material properties are wasted. This also applies to various surface dents and scratches which may appear after careless handling and assembly, often decreasing the expected fatigue life more than twice.

5   Dynamic properties

Most structures are dynamically loaded and so their components may be exposed to unexpected dynamic effects, viz to resonances producing substantially higher loads than considered when designing.   It is therefore sometimes virtually impossible to design a component without changing its dynamic properties, e.g. by introduction of damping (applying a damping layer), changing of its supports (statically determined to statically indetermined), point of energy supply, etc.

6   Fractography

Morphology of fractured surfaces is the best witness of what happened and if properly read it often provides the answer why a specific component got broken. Various marks and features such as "fish eyes", striations, changing roughness, dark and bright fields, (Fig. 5) may indicate where and why the fatigue crack started, at what rate it propagated, whether the corresponding load was high or low, whether the crack initiation was not supported by fretting and/or corrosive environment and many other facts, helping to substantiate the final and unbeatable conclusions.

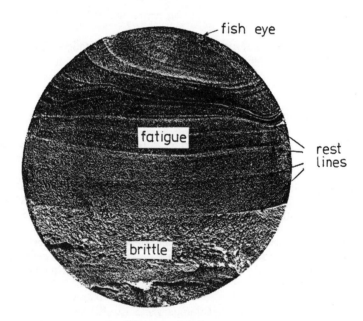

Fig 5: An example of a service fracture with its typical features.

7   Fatigue strength calculation vs fatigue tests

In the literature there are many formulae and computational procedures for estimating the operating fatigue life. However, because fatigue is known to exhibit a wide scatter of lives it is suggested to exploit as many approaches (fatigue damage accumulation hypotheses) as possible, say at least 5 or 6[2]. In this way we get a certain computational "scatter" and "mean value" of the fatigue life which signal where our expectations could lie.

Nevertheless, as pointed out above, the fatigue process is a result of multifactorial effects and it is practically impossible to include all factors in a computational method. So for the critical application the acceptable proof can only be provided by experiments carried out not on one but preferably on more specimens. This suggestion is naturally much more expensive than calculation but it may prevent catastrophes and save human lives.

8   Conclusion

The up-to-date know-how for designing against fatigue cannot unfortunately guarantee 100% success and one should always keep in mind that there is always a reasonable probability that a structure in service will fail, someone will suffer and someone has to pay for it. If one considers the EEC Product Liability Directive 85/374 with its concept of defect in Article 6, viz "*A product is defective when it does not provide the safety which a person is entitled to expect*", then we are provided with a huge arena to play at such interrelated fields as fatigue, legal aspects of service fatigue failures, structural liability, insurance rules, acceptable level of risk and life cycle cost of a structure. It is the privilege and honour of The International Society for Technology, Law and Insurance that it recognizes the social importance of these problems and that it will act as the international coordinator trying to help everyone.

9   References

[1]   Faria, L.: "The Economic Effects of Fracture in Europe", *Commission of the European Communities*, 1991.

[2]   Bily, M.: "Dependability of Mechanical Systems", *Elsevier*, Amsterdam, 1989.

19 FROM A SUB-CRITICAL TO THE CRITICAL CRACK - A COMPLEX DOMAIN IN STRUCTURAL STABILITY

Y. KATZ and A. BUSSIBA
Nuclear Research Center-Negev, Beer-Sheva, Israel

Abstract
The behavior of subcritical defects is probably one of the most striking problems in reliability or structural integrity. Regardless of the exact mechanisms of propagation controlled microcracks, the slow crack growth regime is only partially understood. Nevertheless, this quasi equilibrium stage bears a major role indeed on the acceptable margins against failure or on the improved life-prediction capabilities.

Besides some brief remarks concerning legal aspects and risk analysis, the present study emphasize examples in fatigue and environmental induced cracking as revealed in metallic systems. The main point here to develop some aspects in reliability by describing various intrinsic and extrinsic variables which are imposing complexities on the generic crack stability equation. Generally, by adopting theoretical/experimental interface, critical questions are raised beside suggestions in cultivating complementary activities that promise further progress.

Failure assessment methods still require experimental evaluations and confirmations. The current existing limits as fundamentally explored, should be recognized, believed to initiate partial origins for numerous ill-defined engineering procedures. At least one of the known example is the unsatisfied situation in reliable procedures for life prediction, either in cyclic or in environmental enhancing failure.

1. INTRODUCTION

Here, by way of background, some legal aspects as related to structural failure are outlined, emphasizing current concepts still on regional level. It seems that the key issue lies in an appropriate understanding on what might be expected from intensified legal regulations and how their incorporation might enhance meaningful reliability growth. It is the present understanding that law serves only an essential role in balancing various vital interests providing as such comprehensive benefits. Even these, if based primarily on past experience analyses capabilities and sound expertise input. Established regulations need to be reviewed continuously to allow modifications and the necessary degree of freedom for technological developments. Frequently, tendencies for more of rigid legal approaches arise to mask other deficiencies. Law never claims to exceeds beyond the available wisdom of scientific/technological input, clearly not replacing inherent gaps. In

Structural Failure: Technical, Legal and Insurance Aspects. Edited by H.P. Rossmanith. Published 1996 by E & FN Spon, 2–6 Boundary Row, London SE1 8HN, UK. ISBN: 0 419 20710 4.

fact these are problems along fundamentals in crack physics and other related interfaces. Thus, the current study is guided by realizing the importance of the scientific/technological arena to be responsible for any substantial progress in reliability growth with some well defined and selected assistance by law practice as imposed by society values and norms.

With this in mind, the present study intends to develop some aspects as related to the main conference theme, namely, failure prevention and analysis. Beyond the aforementioned notes on legal technical interface some brief remarks on risk analysis seemed in order. These prior to the elaboration on some specific examples which intend to provide only means to make or clarify few points. Critical questions associated with the subcritical regime allow more of realistic appreciation to the long term objectives in attempting to control fracture events.

Mainly, two cases are described;
(a) Frequency effect observations in fatigue crack propagation rates (FCPR).
(b) Environmental induced slow crack growth in single and polycrystalline materials.

It is well recognized that other important elements are also behind liability aspects but intentially here, only some of the basic limits are mainly emphasized. Although not treated the following elements require further considerations;
(i)   Incident vs. Accident.
(ii)   The role of market constraint in quality. Very powerful factor with important angles and consequences.
(iii) Double standards - military vs. civil high risk products.
(iv)   Statistical aspects on different levels.
(v)   Standards, codes, inspections, specifications and testing.

In the present investigation human aspects are also de-emphasized but well recognized. Even self explained, these aspects which are extremely important need extra care while introducing other categories so essential to be overcome. This in time where international cooperation become almost inevitable.

2. SOME REMARKS ON RISK ANALYSIS

As addressed by Tetelman and Besuner [1] risk may be quantified in terms of hazard index. Following the same spirit but with some modifications;

$$I = P(e) \times P(s)[f(w)] \tag{1}$$

where; I   - Hazard index.
P(e) - Probability for critical event.
P(s) - Probability associated with potential severity.
f(w) - Weight function which implements also social norms.
For the sake of dimensional analysis, eq. (1) can be expressed in terms of global costs.

In fact both P(e) and P(s) are the key elements to be reduced, guided by deterministic and statistical components. This is actually the long term objectives. As shown already [1], the incorporation of a minimum amount of fracture mechanics analysis really provided significant assistance.

A specific case in probability analysis has been addressed by Gamble and Griesbach [2] in the determination of ASME code requirements for pressurized water reactor. This might serve as a good example since the use of fracture mechanics or damage mechanics methods raise the questions of criteria, namely stresses and extension of a existing or postulated flaw.

These avenues but more on a basic level are followed and described, emphasizing the coupled and interactive effects at the crack tip. Needless to say that variations in threshold or critical values, transients in crack propagation rates, reflect directly on risk analysis or improved life prediction methods.

3. EXPERIMENTAL PROCEDURES

3.1 Frequency effect observations

Quenched and aged lean U-Ti alloy was selected consisted of fully lenticular tempered α' martensitic phase with grain size of about 250μm. Chemical analysis was as follows, 0.85 wt% Ti balanced by uranium with impurities of less than 500 ppm.

Standard mechanical properties and fracture toughness parameters were established following ASTM standards. Tests were performed at 295 and 173K. Fatigue tests were conducted on precracked single edge notched, three point bending specimens with dimensions of 55mm in length, 10mm in thickness, 15mm in width and initial crack length of 2mm with load ratio of 0.1. At 295K frequency range between 0.01 to 150Hz was selected in contrast to 0.1 to 10Hz at 173K. Except for the 150Hz, tests were conducted on MTS electro-hydraulic close loop machine. For the high frequency, fully computerized resonant machine was utilized with step down ΔK technique. Crack extension in both cases was monitored continuously by calibrated crack foil gages. Notice that even such observations have to account for interactive effects which need extra experimental care. Particularly closure effects at different temperatures or frequencies and fracture mode variations caused by Ductile-Brittle (D-B) transition, require attention, which are briefly summarized in Table 1.

3.2 Hydrogen effects: Fe-3%Si single crystals

The selected Fe-3%Si single crystal is in the semi-brittle regime with relatively simple and well defined microstructure. With hydrogen interaction an extremely sharp crack tip could be maintained during the subcritical crack extension and the fracture surface morphology exhibited fine scale features that could be resolved on a well defined cleavage plane. These conditions enabled the study of surface energy effects on the local crack extension and on the aspects of cleavage behavior. To further accomplish the programmed goal as related also to hydrogen effects, mini disk shaped compact specimens were utilized about 23mm in diameter with more detailed dimensions and machining methods reported elsewhere [3,4]. Even for the study section of the anisotropic nature of local crack stability alone, particular care was required. Some of the analysis scheme is summarized in Table 2. Crack systems of the two orientations {001} <110> and {001} <010> were examined in order to provide significant results regarding the effective surface energy anisotropy.

Table 1
Variable$_{(*)}$ interactions in frequency or/and temperature effects observations on da/dN$^{(*)}$

Category	Apparent to effective	Possible coupled effects by
Typical sigmoidal$^{(**)}$ curve da/dN vs. ΔK K_{TH} - lower bound	} K_{op} - $K_{TH(eff)}$	T and f (f - frequency, crack tip deformation rate)
K_{Ic} or K_{If} - upper bound	Small closure	
Frequency effects	Need to establish ΔK_{op}	Need to account for isothermal conditions (Adiabatic heat)
Temperature effects	Need to establish ΔK_{op}	Consider the introduction of additional deformation mechanisms (multyslip, interplay of twinning and slip)
K_{eff}	Effects caused by residuals, closure process zone controlling variables might differ in some circumstances (i.e rate of loading effects on damage)	T and f dependent
FCPR		Account for alternative modes static or environmental (internal). Comparative studies required basically similar fracture modes. Aspects in D-B transition

* - Studies must exclude other time dependent factors or embrittlement factors due to temperature.
** - FCPR at the lower shelf or at the intermediate or upper shelf differ dramatically.

Table 2
Analysis regarding anisotropic crack stability [4]

Concepts	Framework	Observations	Analysis	Objectives
		Crack tip surface slip traces	Finite element analysis	Crack tip morphology characterization with emphasis on the activated crystal plasticity features
			Elastic–plastic computer simulation	Micromechanical view with regard to microcrack growth and arrest
Local and micromechanics approach	Internal and external sub-critical flaw growth in single crystals			Driving force with reflection on the local resistance
		Fine scale fracture topography and crack front or void shape	Gibbs–Wulff plots and construction	Semi-equilibrium stabilization trends
				Local resistance aspects with reflection on the local driving force
		Crack initiation and propagation stages	Fracture mechanics considerations	Anisotropic aspects

3.3 Polycrystalline austenitic stainless steel

Here, mainly an experimental program was initiated to explore load inter-action effects activated by single overload. As such, AISI 304 metastable austenitic stainless steel was selected in the form of rolled plates with the following composition (in wt%): Cr-18.6, Ni-9.5, Mn-1.2, Mo-0.23, C-0.05, Si-0.44, Cu-0.25, S-0.016, P-0.033 and Fe-bal.

Load interaction effects were activated by single overload in prefatigued three-point bending specimens. The specimens span length was 100mm, 40mm in width and 19mm in thickness. The range of the overload amplitude strength $q=K_{OL}/K_{max}$ was between 1.35 to 3 at load ratio R~0. After the interrupted fatigue spectra by overloads, FCPR were tracked with and with no hydrogen interactions. The tests were carried out by utilizing a sinusoidal waveform with a frequency of 10Hz. Controlled amplitude range ΔK was kept constant before and after the overload. The overload cycle as applied at 296K resulted mainly in an interactive zone of deformed austenite (γ) while the overload at 77K with subsequent fatigue cracking at 296K resulted in a multiphase zone. Phases of γ, ε' and α', (where ε' and α' are the hcp and the body centered martensitic phases respectively) were confined to the crack tip plastic enclave. The existing phases in the plastic zone were verified in previous studies by X-ray diffraction and Mössbauer spectroscopy [5,6]. Hydrogen charging was performed simultaneously with tension-tension cyclic loading, using a built-in cell containing a 1N H_2SO_4 aqueous solution with current density of 100A/m². As such the FCPR retardation, activated by single overloads was analysed for both cases (as shown in Figure 1 γ as a modified expanded γ phase was also formed).

A scheme summarizing the relevant variables with particular attention to phase stability influences is given in Figure 1.

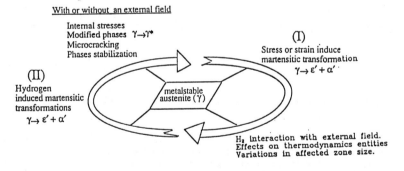

Figure 1. Hydrogen induced cracking - variables interaction and phase stability effects in metastable stainless steel.

4. EXPERIMENTAL RESULTS AND DISCUSSION

4.1 Frequency effect observations

Based on experimental findings, some interesting points are particularly indentified in the U-Ti systems. Although the general trend obeyed FCPR increase as the frequency decrease some significant exceptions were revealed. The role of adiabatic heat and static mode effects were clearly established resulting in dramatic effects. Notice the differences as given in Figure 2 at two levels of the effective stress intensity factor. Here, to mention that these values were determined after closure measurements at 295 and 173K under the corresponding frequencies.

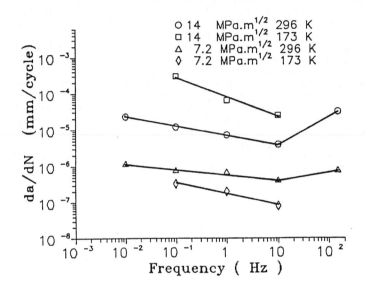

Figure 2. Effect of frequency on FCPR at 296 and 173K at two values of ΔKeff

Moreover, the frequency effect in U-Ti was found to be substantial as compared to AISI 4340 steels at the lower shelf regime [7].

In fact, most of the elements described in Table 1 were actually confirmed experimentally indicating the sharp variations in behavior caused by coupled or interactive factors. In fact, such observations provide two fold purposes;
(a) Fundamental - i.e. to explore static or dislocation dynamic approaches of crack tip processes;
(b) Engineering:
    (i) Evaluation of ultra high frequency tests aimed to establish K_{TH} or FCPR
    (ii) Establishing more realistic constitutive equations.

A good example for the latter is the study of Shimizu et al. and others [8,9] in semiconductors integrated circuits analysis. There, structure reliability was investigated in terms of; stress and fracture analysis, stress simulation in semiconductors fabricating processes, strength evaluation for plastic package design and thermal fatigue strength of solder joints. Thus, the following equation was proposed;

$$\Delta\Gamma = \Delta\tau/G + [\Delta\tau/Ff^m]^{1/n} \qquad (2)$$

where: $\Delta\Gamma$ - damage parameter.
    F - plastic coefficient (strain).
    f - frequency.
    n - strain hardening coefficient depending on temperature (T).
    m - strain rate sensitivity factor where $m = \Psi(T)$.

4.2. Hydrogen effects: Fe-3%Si single crystal

The small near fracture plasticity in the (001) cleavage plane allowed plastic strain distributions determination for subcritical crack growth normal to the crack plane. Some of the novel techniques and the task ahead in confirming analytical or numerical strain distribution for stationary and growing cracks are given elsewhere [10]. These are extremely important activities. Only one finding regarding the crack fronts is invoked indicating again the complex task ahead.

For the {100}<011> crack system activated by a pre-fatigued single notch, a straight local crack front was observed along the <011> direction. However in the case of the {100}<010> crack system, a zig-zag front resulted along two orthogonal <011> direction. A different way to describe it is depicted in Figure 3 for both of the crack systems.

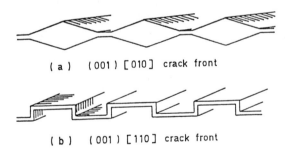

(a)   (001) [010]   crack front

(b)   (001) [110]   crack front

Figure 3. Crack front variations - anisotrpic effect

Beyond the crack front effects multiple nucleation sites along the crack front formed even accentuated with external or internal hydrogen. These multiple sites with "statistical" variations apply to the local stress, microstructure and hydrogen content. Even here the statistical variations may by more deterministic in nature but difficult to analyse at this stage. As adressed by Gerberich and Foecke [11] the interplay of variables is ill-defined and combined procedures of computer simulations, concentration, redistribution and statistical Weibull model allow means to overcome some experimental variations. Following such treatment with dislocation simulations major issue as K_{TH} dependence on temperature and pressure or K_{TH} vs. K_{IC} could be attempted. Generally this was assisted using both Fermi-Dirac statistics, Siverts Law and the concept of relaxed stresses due to a hydrogen-feedback mechanism. Moreover, it was suggested that connectivity exists between single and polycrystalline studies of hydrogen embrittlement of iron-base alloy.

4.3. Polycrystalline austenitic stainless steels

The major findings here revealed the variations in retardation profiles as obtained after a single overload at 77K. This was manifested by subsequent tension-tension fatigue at 296K. At the low temperature the applied thermo-mechanical single overload was far below the M_d resulting in high volume fraction of martensitic phases. In the given microstructures at the shielded crack tip vicinity, substantial differences existed between the subcritical crack extension transients with or with no environmental interactions. Quantitatively, the shielding potential with hydrogen was reduced while significant dependency on the local stress distribution prevailed. In fact fatigue crack propagation rate transients activated by single overloads in 304 metastable austenitic stainless steel involve residual and contact stress effects as process zone-crack tip interactions. The extrinsic shielding potential is associated with intrinsic microstructural properties. In particular residuals due to dilatational stresses become dominant with origins in phase stability aspects.

4.4. Some final remarks

The described cases aimed to realize the ease of transitions while illus-trating variable interactions and the high sensitivity of fracture process to the local crack tip environment. However the challenge to establish firm theoretical basis is an additonal field which needs continuous evaluations.

These latter aspects have been thoroughly addressed by Liebowitz [12]. In this study (particularly in Tables 1-3) the state of the art is overviewed including finite elements analysis, boundary element analysis as the impli-cations of numerical solutions for fracture predictions. However the follow-ing was concluded that fracture mechanics is far from a mature discipline. For example, the issue of unloading requires better understanding as the refinement of local deformation state which is essential and should be encouraged.

5. CONCLUSIONS

(1) Intensified legal regulations as related to reliability growth requires careful evaluation. Law is believed to serve essential role in balancing vital controvarsial interests - not expecting to replace inherent gaps.

(2) Liability remains in the scientific/technological arena and disci-plines such as fracture or damage mechanics might provide breakthroughs even in sound design processes.

(3) Limits should be well recognized, realizing that many sections - both theoretical and experimental are far from being mature.

(4) Experimental confirmation should be encouraged particularly with regard to critical questions. Crack tip field (growing crack) unloading, 3D problems, multiaxial loading (particularly with environmental interactions)

and small cracks. Subcritical crack growth in monotonic, cyclic and environmental induced microcracking still remain key issues.

(5) Case studies, performance tests, inspections, integration of expert systems are important. The recognition of some advanced computational methods as vital for design (numerical analysis) create more of quality attitude.

6. ACKNOWLEDGEMENTS

The authors are grateful to Mr. M. Kupiec and Mr. R. Shefi from the NRCN-Negev for experimental assistance during the present studies.

7. REFERENCES

1   A.S. Tetelman and P.M. Besuner, In Fracture 1977 $\underline{1}$, ICF4, D.M.R.Taplin (ed.), Waterloo, Canada, University of Waterloo Press, (1977) 137.
2   R.M. Gamble and T.J.Griesbach, In ICF7 $\underline{5}$, K. Salama, K. Ravi-Chandar, D.M.R Taplin and P. Rama Rao (eds.), Pergamon Press (1989) 3495.
3   X.F. Chen, J.A. Kozubowski and W.W. Gerberich, Eng. Fracture Mech., $\underline{35}$, (1990) 997.
4   Y. Katz, X. Chen, M.J. Lii, M. Lanxner and W.W. Gerberich, Eng. Fracture Mech., $\underline{41}$, (1992) 541.
5   Y. Katz, A. Bussiba and H. Mathias, In Materials Experimentation and Design in Fatigue, F. Sherratt, J.B. Strugeon and R.A.E. Franborough (eds.), Westbury House, IPL Science and Tech. Press, Guilford, Surrey (1981) 174.
6   H. Mathias, Y. Katz and S. Nadiv, In Metal-Hydrogen Systems, T.N. Veziroglu (ed.), Pergamon Press, Oxford (1982) 225.
7   Y. Katz, A. Bussiba and H. Mathias, In Fatigue at Low Temperatures ASTM STP 857, R.I. Stephens (ed.), ASTM Philadelphia (1985) 191.
8   T. Shimizu, A. Nishimura and S. Kawai, In Fatigue 90 $\underline{2}$, H. Kitagawa and T. Tanaka (eds.), Material and Component Eng. Pub. Ltd., Birmingham, UK (1990) 993.
9   J. Klema et al., Proc. 22nd Int. Reliability Physics Symp. (1984) 1.
10 W.W. Gerberich and S. Chen, In EICM Proceedings Environmental-Induced Cracking of Metals, R.P. Gangloff and M.B. Ives (eds.), National Association of Corrosion Eng., Huston, Texas (1988) 167.
11 W.W. Gerberich and T.J. Foecke, In Hydrogen Effects on Material Behavior, N.R. Moody and A.W. Thompson (esd.), The Minerals, Metals and Materials Society (1990) 687.
12 H. Liebowitz, In ICF7 $\underline{3}$, K. Salama, K. Ravi-Chandar, D.M.R Taplin and P. Rama Rao (eds.), Pergamon Press (1989) 1887.

20 FUNDAMENTALS OF LIMIT-STATE DESIGN OF CONCRETE - RUSSIAN EXPERIENCE

YU. ZAITSEV
Moscow State Open University
and
V. I. SHEVCHENKO
Volgograd Civil Engineering Institute, Russia

The history of reinforced concrete is relatively short. The discovery of reinforced concrete is usually attributed to the work of a Parisian gardener Joseph Monier who made flower pots and orange-tree tubs of cement grout. The pots and tubs often cracked, and for this reason around 1861, Monier came to a good idea to reinforce them with iron mesh. Much to his surprise, they turned out to be exceptionally strong and durable. Still, Monier's discovery might have apparently slipped by unnoticed if the rapidly progerssing civil engineering had not created the need of new building materials.

In 1867, Monier was granted a patent by the French government for the embedding of wire in concrete to add strength to the finished product. Later, he extended his original patent to include the manufacture of reinforced concrete pipes, floors, beams, and bridges. In 1884-86, Monier's patents were bought by some firms in Germany and Austria. By the end of the 19th century, the new building material began its triumphal march across the continents. The World Exhibition of 1900 in Paris came as a culmination in the development of reinforced concrete in the 19th century.

Advances in construction were accompanied by the development of design theory. That time for all the culculations so called *working stress design* (WSD) was used. The growing scale of construction highlighted the drawbacks of elasticity theory treating concrete as an elastic homogenious material. To overcome them, the basic principles of *ultimate-strength design* (USD) were formulated by A. Loleyt in 1931 in the former USSR. This theory suggested that plastic strain in the steel and concrete of a bending beam at failure reached their ultimate values and that determined the critical bending moment.

This approach was backed by a number of experiments and wide theoretical work which gave way to a fundamentally new theory of designing of concrete and reinforced concrete structures. Some time later, the new method was extended to cover members in eccentrical compression. Ultimate-strength theory served as the basis for Soviet standards and specifications

Structural Failure: Technical, Legal and Insurance Aspects. Edited by H.P. Rossmanith. Published 1996 by E & FN Spon, 2–6 Boundary Row, London SE1 8HN, UK. ISBN: 0 419 20710 4.

requiring that reinforced concrete structures should be designed the terms of the breakdown stage.

The second period in the use of reinforced concrete in the former USSR began at the end of the World War II and still continues in the CIS. Duri ng this period, the design layout of structures has undergone a significant change caused by the wide application of structures made entirely of precast members and, in particular, of prestressed members which are manufactured nowadays by almost every plant. Precast skeleton, bearing-wall, and modular multistorey structures have been built, and a theory of their analysis and design has been developed.

In 1955 the limit-state design (LSD) of reinforced concrete members underlying the present-day standards and specifications was developed and put into practice in the former USSR. But before we analyse it, a few words about the structure of concrete and reinforced concrete.

As its name implies, reinforced concrete is composed of plain concrete and reinforcing steel which work together owing to the bond between them. Concrete is   as to say strong in compression and weak in tension. It is, therefore, an easy matter to see that steel bars (having high tensile strength) placed in the tension zone of a concrete member would markedly improve its performance (Fig.1). Reinforced concrete is also a good choice for compessive members because steel is as strong in compression as it is in tension and, being encased in concrete, is practically unable to buckle under load.

Concrete decreases in volume when it is allowed to harden under normal atmospheric conditions (this is known as *shrinkage*) and increases in volume when it hardens in water (this is called *swelling*). Shrinkage is closely related to the physical and chemical processes associated with the hydration of the hardened cement paste. It is most intensive when concrete just begins to harden, becoming less pronounced with time.

Concrete dries nonuniformly throughout its volume, so it shrinks also nonuniformly (the outer layers shrink to a greater extent than the inner layers), resulting in initial shrinkage stresses and strain in the surface layers. However, even if concrete is made to shrink uniformly throughout its volume, shrinkage stresses appear at the surface of the aggregate grains which restrain the free shrinkage of the hardened cement paste. These interface stresses cause s.c. contact or bond cracks between the hardened cement paste and the aggregate grains.

Shrinkage stresses can be reduced by suitably proportioning the concrete mix, by steam and moist curing, and also by providing shrinkage joints in long structures.

So, concrete has a complex conglomerated capillary-porous structure including microcracks and its fracture process during a long period of time can be estimated by methods of fracture mechanics. These methods give the equations, defining conditions of crack initiation and crack propagation under load. Basing on these methods one can establish the conditions for both, crack growth velocity and critical crack size.

Simple methods have been developed for designers in Russia for such culculations. For example, compression test specimens begin to collapse as the previously mentioned microscopic bond-failure cracks merge into large cracks. This takes place when the compressive stress developed in the specimen under load reaches a certain value usually referred to as the point of microcracking. The cracks run mostly parallel or at a small angle to the direction of the compressive force. The sequence of cracking in a concrete prism subjected to a gradually rising compressive load can be simulated using methods of fracture mechanics, which is schematically shown in Fig. 2. Using methods of fracture mechanics it is possible, for instance, to estimate the time to failure of concrete structures under sustained load as the time necessary for a cracks to grow from the initial size to the final critical size.

In design practice initial shinkage stresses and propagation of separate cracks do not enter into the strength analysis of reinforced concrete structures explicitly. Instead, they are taken into account by coefficients accounting for various aspects of concrete strength, and also by reinforcement of concrete structures. Most structures are designed using what is known as the limit-state design (LSD) - it clearly establishes the limit states of structures and, in contrast to the working stress design (WSD) and ultimate strength design (USD) techniques (based on a single factor of safety), sets up a system of design coefficients which guarantee that a structure will not attain such states under the worst from all the possible load combinations and all the possible combinations of strength of the materials.

The design factors employed in the limit-state approach and determined by statistical methods from numerous experiments, include:

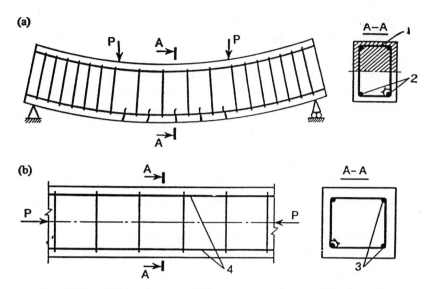

Fig. 1. Distribution of steel in and behaviour of reinforced concrete
members: (*a*) memeber in bending; (*b*) member in   compression
1 - compression zone; *2* - tensile load-bearing   reinforcement;
3,4 - compressive load-bearing reinforcement

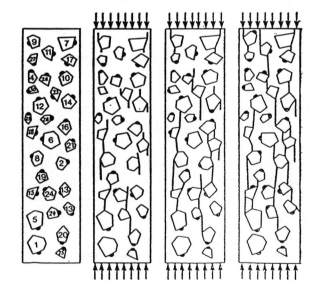

Fig. 2. Sequence of cracking in a concrete
prism subjected to compression

- partial safety factors in terms of load, namely overload factors n taking care of the variability in load and load-combination factors n_c accounting for the real loading conditions;

- partial safety factors in terms of materials, namely strenght variation coefficients v, factors of safety k, service factors m, and reliability indices k_r.

The term "limit state" refers to a boundary condition beyond which a structure of element is considered to be unfit for use either because of inability to support the imposed loads or because of excessive deformation or local failure. The principal limit states to be considered can be divided into 2 parts:

(1) load-carrying capacity, and

(2) serviceability.

The primary objective of design in terms of load-carrying capacity (or design in terms of the first group of limit states) is to prevent a structure or element from a brittle, viscous, or any other failure (if necessary, with allowance for the deformation immediatly before the breakdown). It is also known as design in terms of strength. In addition, it is carried out to be sure that the structure at hand will not buckle under load (this is especially important for thin-walled reinforced concrete structures), tilt or slide along the base (as is the case with retaining walls and similar structures), lose its strength because of fatigue, or collapse when loading is accompained by the unfavourable effects of the environment (alternate freezing and thawing, corrosion, etc.).

Design in terms of serviceability (or design in terms of the second group of limit states) should prevent a structure from exceeding the limit of deformaton (deflection, rotation, warping or vibration), but will also help to keep the width of cracks within the specified limits or even to avoid cracking altogether.

Limit-state analysis and design need to be carried out for any situation that a structure is likely to find itself in during its lifetime, including manufacture, transportation, erection, and service.

Analysis and design in terms of ultimate strength is most important as far as the first group of limit states is concerned. In the general form, the condition of strength for a reinforced concrete member may be written as

$$N_i^{ch} n n_c \leqslant \phi \sum (S, R^{ch}, 1/k_i, m_i), \tag{1}$$

where N_i^{ch} is the internal force caused by the imposed characteristic load, n_i is the overload factor, n_c is the load combination factor, S is the geometrical characteristic of the cross section, R^{ch} is the characteristic strength of the material, k_i is the factor of safety in terms of materials, and m_i is the service factor.

As can be deduced from Eq. (1), a member is said to be adequately strong if the maximum internal force (a bending moment, a shearing force, or an axial force) at the cross section being examined is smaller than (or, at the worst, equal to) the minimum possible loadcarrying capacity of that cross section.

Design Factors [the right-hand side of Eq. (1)]. Relevant standards specify that the main parameter defining the mechanical properties of a material shall be its characteristic strength R^{ch}.

For concrete, three types of characteristic strength are customarily distinquished. These are the ultimate compressive strength determined by testing cube specimens (or the cube crushing strength) R^{ch}. Characterisitc values are established with allowance for the statistical variability of strength and are taken as the least observable ultimate strength:

$$R^{ch} = R \ (1 - k_r v), \tag{2}$$

where $kr = 1.64$ is the reliability index at a minimum level of confidence of 95% (or, equivalently, at a confidence coefficient of 0.95) and v is the coefficient of variation of concrete strength. As it is known from mathematical statistics, the coefficient of variation is defined as the ratio of the standard deviation of a random quantity (concrete strength in our case) to the mean value of that random quantity (in our case, this is the average concrete strength R calculated over a series of tests).

The coefficient of strength variation for precast and cast-in-situ reinforced concrete structures may vary by more than 20%. It depends on the properties and amount of the component materials, the water/cement ratio, curing conditions, test conditions, and other factors.

In Soviet practice, the average coefficient of variation, as found statistically for a large number of concrete product plants and yards, is taken as $v = 0.135$. So, each brand of concrete appears to have a fixed characteristic strength. In manufacture, however, a coefficient of variation in each particular case depends on the process involved. If the actual coefficient of variation turns out to be around the standard 0.135, the requierd average strength will correspond to the brand number specified. If it is below the standars value, the average strength of the concrete may be taken lower than the requerd design number to save cement. If, on the other hand, the manufacturing process involved is such that the actual coefficient of variation exceeds 0.135, the average concrete strength will have to be uprated as compared with the design brand number, which will inevitably entail overexpenditure of cement.

Service Factors. Some conditions which cannot be explicitly taken into consideration in design are accounted for by multiplying the design strength of the concrete or that of the steel by a service factor m_c or m_s. These factors can reduce or increase (as the case may be) the respective design strength of the materials (see Tables I and II).

Let us dwell upon the service factor m_{c1} which accounts for the probable duration of loading. It is known from experience that, when subjected to a sustained load, concrete may break down at a stress which is a only 75% to 85% of its ultimate strength (provided the strength of the concrete does not build up during the loading).

Another service factor, m_{c3} less or equal to 1, is applied where a structure is to resist alternate freezing and thawing in a water-saturated condition, which, in some situations, may cause the concrete to fail.

The service factor m_{c4} more than 1 is applied where prestressed srtructures are analysed for strength during prestressing. It takes care of the short duration of this loading and also of the fact that the failure of a member during the transfer of the prestress is fraught with less serious consequences than in service.

The service factor $m_{c5} = 0.9$ refers to plain concrete structures which, according to a relevant standard, are assumed to owe their strength in service to the concrete alone.

The service factor m_{c7} accounts for the reduced strength of the upper portion of cast-in-situ concrete where layer of concreting is more than 1.5 m deep.

The service factor $m_{c8} = 0.85$ takes care of the likely negative (often considerable) effect of cavities and similar defects on the load-carrying capacity of columns having a small cross-sectional area (less than 30 x 30 cm) where these defects may occupy a considerable portion of the cross section.

Conversely, the service factor $m_{c9} = 1.15$ allows for the increased strength of narrow joints. It is governed by the resistance to lateral expansion, offered at a joint by the ends of the adjoining members.

Table 1. Service Factors for Concrete

Conditions accounted for by service factors	Symbol	Value	Design strength to be multiplied by service factor
Duration of loading:			
(a) for deal, long- and short-time live loads (except loads of relatively short total duration*), and also special loads due to deformation of subsident and other soils:	m_{c1}		R_{pr}, R_{tcn}
- under service conditions favourable for strength build-up**		1	
- in other cases		0.85	
(b) for dead and live loads of short total duration, and also special loads not included in case "a"		1.1	
Multiple repeated loading	m_{c2}	0.45-1	R_{pr}, R_{tcn}
Alternate freezing and thawing	m_{c3}	0.7-1	R_{pr}
Prestressing of members reinforced with:			
- wire	m_{c4}	1.1	R_{pr}
- bars		1.2	
Plain concrete structures	m_{c5}	0.9	R_{pr}, R_{tcn}
Concreting of members in vertical position, with concreting layer deeper than 1.5 m	m_{c7}	0.85	R_{pr}
Cast-in-situ plain concrete piers and reinforced concrete columns with longer side of section less than 30 cm	m_{c8}	0.85	R_{pr}
Joints between prefab members, with joint width smaller than one-fifth of smallest section dimension and narrower than 10 cm	m_{c9}	1.15	R_{pr}
Autoclaving	m_{c10}	0.85	R_{pr}
Exposure to direct solar radiation (in regions with hot arid climate)	m_{c11}	0.85	R_{pr}, R_{tcn}

* For example, crane, wind, transit, and similar loads.

** Hardening under water, in damp soil, or at an ambient humidity of more than 75%.

Table 2. Service Factors for Reinforcing Steel

Type and class of steel	Conditions accounted for by service factors	symbol	Service factors value
Class A-I, A-II, A-III, A-IV, B-I, Bp-II, and K-7 longitudinal and transverse steel	Multiple repeated loading	m_{s1}	0.3-1.0
Class A-I, A-II, and A-III longitudinal and transverse steel	Welded joints in reinforcement of members subjected to multiple repeated loading	m_{s2}	0.2-1.0 (to be applied in conjunction with m_{s1})
Longitudinal reinforcement of all classes	Loading within transmission length of prestresses steel without anchorage and within anchorage length of non-prestressed steel	m_{s3}	l_x/l_{tr} for prestressed steel l_x/l_{an} for non-prestressed steel
Class A-IV, At-IV, A-V, At-VI, B-II, Bp-II, and K-7 longitudinal tensile steel	Loading of high-strength steel beyond proof stress	m_{s4}	1 [as in Eq. (IX.1) in Chap. IX]

Notation: l_x = distance from the beginning of the stress transmission zone to the cross section being examined
l_{tr} = transmission length
l_{an} = anchorage length

The service factor $m_{c10} = 0.85$ accounts for the increased brittliness of the concrete in members cured by autoclaving.

Finally, the service factor m_{c11} applies to structures lacking adequate protection against intensive solar radiation and erected in localities with a hot, dry climate where the sun rays may heat the concrete to more than 50°C, thereby reducing its strength.

For structures subjected to multiple repeated loading, the factor m_{c1} is introduced in strength analysis, and the factor m_{c2}, in fatigue analysis.

As already mentioned the objective of design in terms *of the second group* of limit states is to prevent a structure from exceeding the limit of deformation (deflection, rotation, warping, or vibration) and also to keep the width of cracks within the specified limits or even to avoid cracking altogether. Crack-resistance or crack-width analysis is carried out considering the category of crack-resistance requirements that a given structure is to meet and also with allowance for the sensitivity of the chosen reinforcement to corrosive environments.

The crack resistance of a reinforced concrete structure is defined as its ability to resist cracking or crack opening in a state of stress. Depending on the exposure conditions, the requirements for the crack resistance may be divided into three categories:

- Category one, no cracks are allowed;

- Category two, short-time opening of limited-width cracks $(a_{cr,sht})$ is allowed provided that they close tightly after the load has been removed;

- Category three, short- and long-time opening of limited-width cracks $(a_{cr,\ sht}$ and $a_{cr,\ lt})$ is allowed.

The term "short-time" refers to the opening of cracks under the action of dead and short- and long-time live loads (that is, under all possible types of load applied simultaneously). The term "long-time" refers to the opening of cracks under the action of dead and long-time live loads.

Sometimes, appropriate standards require that members should be checked for closure of normal and inclined cracks. Above all, this refers to members which are to meet Category two crack-resistance requirements and which may contain limited-width cracks (due to dead and long- and short-time live loads).

Author index

Subject index